Ethics and human values in engineering practices

Ethics and human values in engineering practices

Dr. Subrata Das

WOODHEAD PUBLISHING INDIA PVT LTD

New Delhi

Published by Woodhead Publishing India Pvt. Ltd.
Woodhead Publishing India Pvt. Ltd.,
303, Vardaan House, 7/28, Ansari Road,
Daryaganj, New Delhi - 110002, India
www.woodheadpublishingindia.com

First published 2022, Woodhead Publishing India Pvt. Ltd.
© Woodhead Publishing India Pvt. Ltd., 2022

Woodhead Publishing India Pvt. Ltd. ISBN : 978-81-954048-3-4
Woodhead Publishing India Pvt. Ltd. e-ISBN : 978-81-954048-1-0

Typeset by Bhumi Graphics, New Delhi
Printed and bound by Atlantic Publishers & Distribution (P) Ltd

Contents

Preface   *xi*

1. Human values   **1**

   1.1   Introduction   1

   1.2   Difference between morals and values   3

   1.3   Difference between ethics and values   4

   1.4   Difference between beliefs and values   4

   1.5   Relationship between values, morals and ethics   4

   1.6   Comparison between ethics and morals   5

   1.7   Conflicts between ethics and morals   6

   1.8   Honesty   7

   1.9   Integrity   7

   1.10   Values   8

   1.11   Evolution of human values   10

   1.12   Work ethics   11

   1.13   Civic virtue   12

   1.14   Respect for others   13

   1.15   Living peacefully   13

   1.16   Caring and sharing   14

   1.17   Self-confidence   15

   1.18   Courage   16

   1.19   Co-operation   17

   1.20   Commitment   18

   1.21   Empathy   19

   1.22   Spirituality in business   19

   1.23   Service learning   20

2. Engineering ethics   **23**

   2.1   Overview of engineering ethics   23

   2.2   Importance   23

2.3   Scope   24

2.4   Variety of moral issues   24

2.5   Types of inquiry   27

2.6   Accepting and sharing responsibility   28

2.7   Ethical dilemmas   29

2.8   Moral autonomy   30

2.9   Kohlberg's theory   30

2.10 Gilligan's theory   32

2.11 Gilligan's view of Kohlberg   33

2.12 Consensus and controversy   34

3. Professionalism   **36**

3.1   Profession and professionalism   36

3.2   Right action theories   40

3.3   Uses of ethical theories and their limitations   45

3.4   Senses of corporate responsibility   46

3.5   Codes of ethics   46

3.6   NSPE (National Society of Professional Engineers)   48
      code of ethics

3.7   The Institute of Electrical and Electronics Engineers   51
      (IEEE) code of ethics

3.8   Institution of Engineers (India) code of ethics   53

4. Engineering as social experimentation   **56**

4.1   Engineering as experimentation   56

4.2   Conscientiousness   58

4.3   Broad outlook   58

4.4   Moral sovereignty   59

4.5   Accountability   59

4.6   Balanced outlook on law   60

4.7   Significance of law   60

4.8   Cautious optimism   61

4.9   Safety and risk   61

4.10 The Challenger case study   64
4.11 Bhopal gas tragedy   65
4.12 The Three Mile Island (TMI)   67
4.13 Chernobyl disaster   69

5. Workplace rights and responsibilities   73
5.1   Introduction   73
5.2   Professional rights   74
5.3   Employee rights   74
5.4   Right to privacy   75
5.5   Right to choose outside activities   75
5.6   Right to due process from the employer   75
5.7   Teamwork   76
5.8   Ethical corporate climate   76
5.9   Collegiality   78
5.10 Loyalty   79
5.11 Managing conflict/conflict of interest   80
5.12 Respect for authority   81
5.13 Collective bargaining   83
5.14 Unionism and professionalism   83
5.15 Confidentiality   84
5.16 Occupational crime   85

6. Global issues   87
6.1   Multinational corporations   87
6.2   Technology transfer   88
6.3   Appropriate technology   88
6.4   International human rights   88
6.5   Promoting morally just measures   89
6.6   Environmental ethics   91
6.7   Role of engineers in sustainability   91
6.8   Engineers responsibilities towards environment   93

6.9   Engineering, ecology and economics   95
6.10   Computer ethics and internet   99
6.11   Weapons development   103
6.12   Engineers as managers   104
6.13   Solving conflicts   104
6.14   Consulting engineers   105
6.15   Engineers as expert witness   107
6.16   Engineers as advisors   109
6.17   Moral leadership   110

Index   **113**

Preface

Professional engineers must have ethics to make decisions that should be self-reliant and autonomous. They should be morally committed and equipped to tackle ethical dilemmas they face and to hold paramount the safety, health and welfare of the public. The actions of an engineer must be like a true professional.

Engineering ethics is more than mere knowledge and skills, and that the final goal of engineering ethics is to foster qualities and abilities that enable engineers to make self-reliant/autonomous decisions and actions as professionals. The objective of this book is to introduce the readers to the ethical concepts that lead to resolving moral issues in engineering, understanding of ethics and responsibility of engineers as professionals, ability to make ethical judgments and solve problems, attitude required and values shared by engineers to interface social, technological and natural environments.

Humans have the unique ability to define their identity, choose their values and establish their beliefs. All three of these directly influence a person's behavior. Such basic human values are discussed in Chap. 1. Engineering ethics are covered in Chap. 2, which deals with (a) understanding the moral values that ought to guide the engineering profession, (b) resolve the moral issues in the profession and (c) justify the moral judgment concerning the profession. It is intended to develop a set of beliefs, attitudes and habits that engineers should display concerning morality. Professional societies promulgate the codes of conduct to regulate the professionals against their abuse or any unethical decisions and actions (impartiality, responsibility) affecting the individuals or groups or the society. Chapter 3 deals with professionalism with require certain attitudes and typical qualities that are expected from a professional. All products of technology present some potential dangers, and thus engineering is an inherently risky activity. To underscore this fact and help in exploring its ethical implications, it is suggested that engineering should be viewed as an experimental process. It is not, of course, an experiment conducted solely in a laboratory under controlled conditions. Rather, it is an experiment on a social scale involving human subjects. Engineering as social experimentation is discussed in Chap. 4. Human rights are defined as moral entitlements that place obligations on other people to treat one with dignity and respect. Organisations and engineers are to be familiar with the minimum provisions under human rights so that the engineers and organizations constitute a firm base for understanding and productivity. Employee rights are the moral and legal rights obtained with employee status and the provisions are underlying in professional rights, basic human rights, institutional rights and non-contractual employee rights. These are explained in Chap. 5. Globalization means integration of countries through commerce, transfer of technology, and exchange of information and culture. The increasing international flow of capital, technology, trade and people influenced to change in the nature of local organizations, governments and people that resulted in social changes and development. The issues such as multinational organizations, computers, internet functions, military development and environmental ethics have assumed greater importance for their very sustenance and progress for the engineers. Those subjects are explained in Chap. 6.

The book is aimed at industry professionals, retailers, factory heads, buying offices, academicians and students intending to join any industry in the areas of management, production, research and development and quality assurance. To produce and deliver better output, it is necessary to adhere to the right human values, professionalism, laws and regulations applicable in the industry both national and international level. Thus, the emphasis throughout the book is on. engineering and environmental ethics, moral values, human rights and global issues.

Date: 18 November 2019
Place: Sathyamangalam

Dr. Subrata Das
Professor (Fashion Technology)
Bannari Amman Institute of Technology,
Sathyamangalam, Erode District,
Tamil Nadu – 638401

Human values

Abstract: Humans have the unique ability to define their identity, choose their values and establish their beliefs. All three of these directly influence a person's behaviour. A harmonious development and relationship of all these values help us to grow as true human beings. Our technical education should be based on ideas drawn from both classical value tradition and modern humanistic thoughts. The main objective of such an education would teach us how to be a good human being, how to live a good life and how to help create a good society. The academic enquiry pursuing these varied dimensions of human life can be generally described as human values. Value education is particularly important in engineering, management, medicine and law. It is important to understand the impact of their chosen profession on the society at large in the wider social and human context. Creation of value concept in the appropriate climate will encourage the emergence of good human beings, a band of worthy as well as socially responsible professionals and will eventually lead to the creation of a good society. The chapter first discusses relationships, differences and conflicts of values, morals and ethics. The chapter then discusses about honesty, integrity, core human values, work ethics, civic virtue, respect for others, living peacefully, caring and sharing, self-confidence, courage, co-operation, commitment, empathy, spirituality in business, and service learning.

Keywords: human values; morals, ethics;integrity; honesty; empathy

1.1   Introduction

Human values are imperative to understand the moral values that should guide the engineering profession, resolve the moral issues in the profession, and justify the moral judgment concerning the profession. It is intended to develop a set of beliefs, attitudes, and habits that engineers should display concerning morality. To improve the cognitive skills of professionals, moral awareness, moral reasoning, moral imagination and moral communication are given weight age to express and support one's views to others. To act in morally desirable ways, towards moral commitment and responsible conduct, moral reasonableness, respect for persons beside oneself, tolerance of diversity, moral hope and integrity always sets professional life and personal convictions in focus. Essential elements of human values include but are not limited to morals, values, ethics, integrity, service learning, virtues, respect for others, leaving peacefully, caring and sharing, honesty, self-confidence, courage, co-operation, commitment, empathy and spirituality.

Values are the guiding principles of our lives. They are essential for positive human behaviour and actions in our daily lives. They are formed based on interests, choices, needs, desires and preferences. They have played an important role not only in sociology but also in psychology, anthropology and related disciplines [1].

We encounter several circumstances every day that test our patience, our character and peace of mind. We have to make tough decisions each day. What guide us in these circumstances are our values. Our values serve as markers to tell if life is heading in the right direction. When our actions and words are aligned with our values, life feels good and we feel content, confident and satisfied. But when our behaviour does not concide with our values, we sense an uneasiness that grows inside us. This uncomfortable feeling tells us that everything is not good right now. We feel out-of-sorts. These feelings can be a source of anxiety and unhappiness. We need value in our lives to:

- Guide us on the right path.
- Learn the importance of certainty, goodness and beauty.
- Give direction to life and bring joy.
- Learn satisfaction towards life.
- Attain peace in life.
- Develop character.
- Preserve our culture and heritage.
- Bring changes in behaviour towards positive thoughts.
- Promote peace and harmony in the society.

The term intrinsic means 'in itself' or 'for its own sake'. Intrinsic values are those values that have eternal property without any reference to any end. For example, happiness or peace or joy or truth is an intrinsic value. Extrinsic values are those whose property or value depends on how much it generates the intrinsic values. Having a family is an extrinsic value because its value depends on how much happiness or joy it creates. Basic human values refer to those values which are at the core of being human. The values which are considered basic inherent values in humans include truth, honesty, loyalty, love, peace, etc. because they bring out the fundamental goodness of human beings and society at large. Further, since these values are unifying in nature and cut across individual's social, cultural, religious and sectarian interests;

they are also considered universal, timeless and eternal applying to all human beings. Values and norms are different. Norm refers to a relatively specific behaviour as per social customs and it is obligatory. On the other hand, values are a matter of choice. For example, honesty cannot be a norm because it may not be chosen to be followed. Further, once a particular value is internalized by an individual, it becomes a norm for him/her for making decisions, judgments, preferences and choices.

1.2   Difference between morals and values

Morals are taught by the society to the individual while values can be cultivated from within. Morals act as motivation for leading a good life, while values act as intuition. Further, while morals are deep-rooted, values may keep changing from time to time and as per needs. There are six main features of values as per the value theory of Schwartz [2,3]. Values are beliefs linked to affect, which implies that when values are activated, they become infused with feeling. For example, people for whom independence is an important value become aroused if their independence is threatened, despair when they are helpless to protect it and are happy when they can enjoy it. Values refer to desirable goals that motivate action. It indicates that people for whom social order, justice and helpfulness are important values are motivated to pursue these goals. Values transcend specific actions and situations. Obedience and honesty, for example, are values that may be relevant at work or in school, in sports, business, and politics, with family, friends, or strangers. This feature distinguishes values from narrower concepts like norms and attitudes that usually refer to specific actions, objects or situations. Values serve as standards or criteria and guide the selection or evaluation of actions, policies, people and events. People decide what is good or bad, justified or illegitimate, worth doing or avoiding, based on possible consequences for their cherished values. But the impact of values in everyday decisions is rarely conscious. Values enter awareness when the actions or judgments one is considering have conflicting implications for different values one cherishes. Values are ordered by importance relative to one another. People's values form an ordered system of value priorities that characterize them as individuals. Do they attribute more importance to achievement or justice, to novelty or tradition? This hierarchical feature also distinguishes values from norms and attitudes. The relative importance of multiple values guides to action. Any attitude or behaviour typically has implications for more than one value. For example, attending prayer might express and promote tradition, conformity, and security values at the expense of hedonism and stimulation values. The trade-off among relevant, competing values is what guides attitudes and behaviours. Values contribute to action to the extent that they are relevant in the context (hence likely to be activated) and important to the actor.

1.3   Difference between ethics and values

Ethics is a branch of philosophy that is used to study ideal human behaviour and ideal ways of being. Social standards evaluate what is ethical and unethical and vary from person to person. Values are the embodiment of what an individual stands for, and they are basis for the behaviour that forms the basis for ethics. Both ethics and values are situational and changeable in relevant circumstances.

1.4   Difference between beliefs and values

A belief is an internal feeling that something is true, even though that belief may be unproven or irrational. For example, I believe that if I see a black cat crossing the road, it indicates bad luck. On the other hand, a value is a measure of the worth or importance a person attaches to something. Our values are often reflected in the way we live our lives, for instance, we value freedom of speech or we value our families.

All of us have a constant internal battle between our beliefs and values. Sometimes, we mistake our beliefs as values or vice versa. Beliefs are internal, while values are external. This implies that we can pick up a value from an external source or experience, person or thing and start living with that value inculcated in us. But belief is an internal energy that is created on what we absorb and then it builds itself within us further creating our thoughts, words and actions.

Our beliefs create thoughts; thoughts create emotions; emotions create actions of positive values or negative values which depend on the quality of the belief itself. These then become internal values.

1.5   Relationship between values, morals and ethics

The moral values in our lives hold great importance from the point of personal, social and spiritual development. Values, morals and ethics are inextricably tied together. The preservation of human life is the ultimate value, a pillar of ethics and the foundation of all morality [4].

Values are what we learn from childhood; the stuff we acquired from our parents and immediate surroundings. Values are the motive power behind purposeful action. Moral values are meant for making the quest to find the higher self easier. Many amongst us may find it difficult to follow values, such as truthfulness, honesty, forgiveness in our lives because we have not perceived the subtle gains that come to us by following these values. Alternatively, it may be due to the fact that we are careless to realise the importance of values in life [4].

Ethics, on the other hand, are how we do behave in the face of difficult situations that test our morality. Ethics is the code or principle on which one's character depends. Ethics and character are closely related. Values are essential to ethics to develop at an early age and can be instrumental to building character [4].

Whereas, morals are the intrinsic beliefs developed from the value systems of how we 'should' behave in any given situation. Moral values are the standards of good and evil, which govern an individual's behaviour and choices [4].

Ethics and morals relate to 'right' and 'wrong' conduct. While they are sometimes used interchangeably, they are different: ethicsrefer to rules provided by an external source, for example, codes of conduct in workplaces or principles in religions. Morals refer to an individual's principles regarding right and wrong. A comparison between ethics and morals is shown in the below chart [5].

1.6   Comparison between ethics and morals [5]

Statement	Ethics	Morals
What are they?	The rules of conduct are recognized for a particular class of human actions or a particular group or culture	Principles or habits for right or wrong conduct. While morals also prescribe dos and don'ts, morality is ultimately a personal compass of right and wrong
Where do they come from?	Social system —external	Individual —internal
Why we do it?	Because society says it is the right thing to do	Because we believe in something being right or wrong
Flexibility	Ethics are dependent on others for definition. They tend to be consistent within a certain context but can vary between contexts	Usually consistent, although can change if an individual's beliefs change

Contd...

Contd...

Statement	Ethics	Morals
The 'Gray'	A person strictly following ethical principles may not have any morals at all. Likewise, one could violate ethical principles within a given system of rules to maintain moral integrity	Amoral person although perhaps bound by a higher covenant, may choose to follow a code of ethics as it would apply to a system. 'Make it fit'
Origin	Greek word 'ethos' meaning 'character'	Latin word 'mos' meaning 'custom'
Acceptability	Ethics are governed by professional and legal guidelines within a particular time and place	Morality transcends cultural norms

Ethics are external standards that are provided by institutions, groups or cultures to which an individual belongs. For example, lawyers, policemen and doctors all have to follow an ethical code laid down by their profession, regardless of their feelings or preferences. Ethics can also be considered a social system or a framework for acceptable behaviour. Morals are also influenced by culture or society, but they are personal principles created and upheld by individuals themselves. Ethics are very consistent within a certain context but can vary greatly between contexts. For example, the ethics of the medical profession in the 21st century are generally consistent and do not change from hospital to hospital, but they are different from the ethics of the 21st-century legal profession. An individual's moral code is usually unchanging and consistent across all contexts, but it is also possible for certain events to radically change an individual's personal beliefs and values.

1.7   Conflicts between ethics and morals

One professional example of ethics conflicting with morals is the work of a defense attorney. A lawyer's morals may tell him/her that murder is reprehensible and that murderers should be punished, but his/her ethics as a professional lawyer, require to defend the client to the best of him/her abilities, even if he/she knows that the client is guilty. Another example can be found in the medical field. In most parts of the world, a doctor may not euthanize a patient, even at the patient's request, as per ethical standards for health professionals. However, the same doctor may personally believe in a patient's right to die, as per the doctor's morality. Much of the confusion between these two words can be traced back to their origins. For example, the word 'ethic' comes from old French (etique), late Latin (ethica) and Greek (ethos) and referred to customs or moral philosophies. 'Morals' comes from late Latin's morals, which referred to appropriate behaviour and manners in society. So, the two have very similar, if not synonymous, meanings originally. The morality and ethics of the individual have been philosophically studied for well over a thousand years. The idea of ethics being principles that are set and applied to a group (not necessarily focused on the individual) is relatively new, though, primarily dating back to the 1600s. The distinction between ethics and morals is particularly important for philosophical ethicists.

1.8 Honesty

As per Chanakya [6], a person should not be too honest. Straight trees are cut first and honest people are screwed first. However, honesty is a virtue, and it is exhibited in two aspects namely, (a) truthfulness and (b) trustworthiness [7].

Truthfulness is to face the responsibilities upon telling truth. One should keep one's word or promise. By admitting one's mistake committed (one needs the courage to do that!), it is easy to fix them. Reliable engineering judgment, maintenance of the truth, defending the truth, and communicating the truth, only when it does 'good' to others, are some of the reflections of truthfulness.

Trustworthiness is maintaining the integrity and taking responsibility for personal performance. People abide by the law and live by mutual trust. They play the right way to win, according to the laws or rules (legally and morally). They build trust through reliability and authenticity. They admit their own mistakes and confront unethical actions in others and take a tough and principled stand, even if unpopular.

Honesty is mirrored in many ways. The common reflections are:
(a) Beliefs (intellectual honesty).
(b) Communication (writing and speech).
(c) Decisions (ideas, discretion).
(d) Actions (means, timing, place and the goals).
(e) Intended and unintended results achieved.

1.9 Integrity

Integrity stems from the Latin word 'integer' which means whole and complete. So integrity requires an inner sense of 'wholeness' and consistency of character. Integrity is defined as the unity of thought, word and deed (honesty) and open-mindedness. It includes the capacity to communicate factual information so that others can make well-informed decisions. It yields the person's 'peace of mind', and hence adds strength and consistency in character, decisions and actions. This paves the way to one's success. It is one of the self-direction virtues. It enthuses people not only to execute a job well but to achieve excellence in performance. It helps them to own the responsibility and earn self-respect and recognition by doing the job.

Integrity is the quality of being honest and having strong moral principles or moral uprightness. It is generally a personal choice to hold oneself to consistent moral and ethical standards [8]. *Integrity is a personal choice, an uncompromising and predictably consistent commitment to honour moral, ethical, spiritual and artistic values and principles.*

1.10   Values

Humans have the unique ability to define their identity, choose their values and establish their beliefs. All three of these directly influence a person's behaviour. People have gone to great lengths to demonstrate the validity of their beliefs, including war and sacrificing their own life! Conversely, people are not-motivated to support or validate the beliefs of another, when those beliefs are contrary to their own. People will act congruently with their values or what they deem to be important.

A value is defined as 'a principle that promotes well-being or prevents harm' Another definition is: 'Values are our guidelines for our success – our paradigm about what is acceptable.'Personal values are defined as: 'Emotional beliefs in principles regarded as particularly favourable or important for the individual.'Our values associate emotions with our experiences and guide our choices, decisions, and actions.

1.10.1   Types of values

The five core human values are: (1) right conduct, (2) peace, (3) truth, (4) love and (5) non-violence [9].

1.10.1.1   Right conduct

True right conduct is speaking and acting on the truth that emerges from the heart, the source of human conscience and human values. When we engage in the right action, we treat others the way we wish to be treated, with respect, kindness, compassion, with an understanding and appreciation of the unity of all life.

Values related to right conduct are:(a) self-help skills: care of possessions, diet, hygiene, modesty, posture, self-reliance and tidy appearance; (b) social skills: good behaviour, good manners, good relationships, helpfulness, no wastage and good environment and(c) ethical skills: code of conduct, courage, dependability, duty, efficiency,ingenuity, initiative, perseverance, punctuality, resourcefulness, respect for all and responsibility.

1.10.1.2 Peace

Absolute peace manifests as inner mental calm, and the ability to maintain equanimity in all situations. When we feel peace within ourselves we will naturally feel peaceful towards others. Values related to peace are attention, calmness, concentration, contentment, dignity,discipline, equality, equanimity, faithfulness, focus, gratitude, happiness, harmony, humility,inner silence, optimism, patience, reflection, satisfaction, self-acceptance, self-confidence,self-control, self-discipline, self-esteem, self-respect, sense control, tolerance and understanding.

1.10.1.3 Truth

The highest truth is absolute, changeless in the past, present and future, true at all times and in all places. Truth is not relative, changing according to our perceptions and circumstances. Truth manifests as being truthful, honest and sincere, acting with integrity according to the dictates of our conscience.

Values related to truth are accuracy, curiosity, discernment, fairness, fearlessness, honesty,integrity (unity of thought, word and deed), intuition, justice, optimism, purity, quest for knowledge, reason, self-analysis, sincerity, spirit of enquiry, synthesis, trust, truthfulness and determination.

1.10.1.4 Love

The highest love is selfless love. It is the love that is unconditional, without attachment, not expecting anything in return. All actions emanate from the heart. Love is the force behind the other human values and our actions:

Values related to love are: acceptance, affection, care, compassion, consideration, dedication, devotion, empathy, forbearance, forgiveness, friendship, generosity, gentleness,humanness, interdependence, kindness, patience, patriotism, reverence, sacrifice, selflessness,service, sharing, sympathy, thoughtfulness, tolerance and trust

1.10.1.5 Non-violence

Avoiding causing harm to anyone or anything in our thoughts, words, and deeds is called non-violence. Non-violence allows us to appreciate diversity, cultivate tolerance and recognize the unity of all beings and respect for all life.

Values related to non-violence are: (a) psychological: benevolence, compassion, concern for others, consideration,forbearance, forgiveness, manners, happiness, loyalty, morality, and universal love; (b) social: appreciation of other cultures and religions, brotherhood, care of the environment,citizenship, equality, harmlessness, national awareness, perseverance, respect for property and social justice.

Perseverance is defined as persistence, determination, resolution, tenacity, dedication,commitment, constancy, steadfastness, stamina, endurance and indefatigability. To persevere is described as to continue, carry on, stick at it (informal), keep going, persist, plug away, (informal), remain,stand firm, stand fast, hold on and hang on. Perseverance builds character.

Accuracy means freedom from mistake or error; conformity to truth or a standard or model and exactness. Accuracy is defined as correctness, exactness, authenticity, truth, veracity, closeness to truth (true value) and carefulness. The value of accuracy embraces a large area and has many implications. Engineers are encouraged to demonstrate accuracy in their behaviour through the medium of praise and other incentives. Accuracy includes telling the truth, not exaggerating and taking care of one's work.

Discernment means discrimination, perception, penetration and insight. Discernment means the power to see what is not obvious to the average mind. It stresses accuracy, especially in reading character or motives. Discrimination stresses the power to distinguish or select what is true or genuine lyexcellent. Perception implies quick and often sympathetic discernment, as of shades of feelings. Penetration implies a searching mind that goes beyond what is obvious or superficial. Insight suggests a depth of discernment.

1.11   Evolution of human values

The human values evolve because of the following factors [8]:

1. The impact of norms of the society on the fulfilment of the individual's needs or desires.
2. Developed or modified by one's awareness, choice and judgment in fulfilling the needs.
3. By the teachings and practice of preceptors (Gurus) or saviors or religious leaders.
4. Fostered or modified by social leaders, rulers of the kingdom and by law (government).

1.12   Work ethics

Work ethics is defined as a set of attitudes concerned with the value of work, which forms the motivational orientation. The 'work ethics' is aimed at ensuring the economy (get job, create wealth, earn salary), productivity (wealth, profit), safety (at workplace), health and hygiene (working conditions), privacy (raise a family), security (permanence against contractual, pension,and retirement benefits), cultural and social development (leisure, hobby and happiness), welfare (social work), environment (anti-pollution activities) and offer opportunities for all, according to their abilities, but without discrimination.

Execution of work (job), is not for monetary considerations only. Human beings believe that it is good to work. Work is good for the body and mind. It promotes self-respect, self-esteem, well for the family, and obligation to society and allows the world to prosper. Work lays a moral and meaningful foundation for life. Therefore, work ethics affirms that work per se is worthy, admirable and valuable at personal and social levels. It improves the quality of life and makes life purposeful, successful and happy.

By work ethics, duties to the self, family, society, and nation are fulfilled. The rights of the individuals are respected and nourished. Values and virtues are cultivated and enjoyed by all human beings. Further, the quality of life is improved and the environment protected. On the other hand, unemployment and under-employment lead to frustration, social tensions and occasional militancy. For a developing economy and society, one needs to promote work ethics at all levels to flourish as a developed nation.

Industry and society are two systems that interact with each other and are interdependent. Society requires an industry/business system that provides manufacturing, distribution, and consumption activities. It needs investment (capital input), labour (input), supply (raw materials), production (industries, business organizations), marketing and distribution (transport) and consumption (public, customer). A lot of transactions (and interactions) between these sub-systems involving people are needed for the welfare of society. It is here, the work ethics plays an essential role.

By work ethics, duties to the self, family, society and nation are fulfilled [7]. The rights of the individuals are respected and nourished. Values and virtues are cultivated and enjoyed by all human beings. Further, the quality of life is improved and the environment protected. On the other hand, unemployment and under-employment lead to frustration, social tensions and occasional militancy. For a developing economy and society, like ours, we need to promote work ethics, at all levels, to flourish as a developed nation.

1.13   Civic virtue

Virtues are positive and preferred values. Virtues are desirable attitudes or character traits, motives and emotions that enable us to be successful and to act in ways that develop our highest potential. They energize and enable us to pursue the ideals that we have adopted. Various examples of virtues are honesty, courage, compassion, generosity, fidelity, integrity, fairness, transparency, self-control, prudence.

Virtues are tendencies that include, solving problems through peaceful and constructive means and follow the path of the golden mean between the extremes of 'excess and deficiency'. They are like habits, once acquired, they become characteristics of a person. Moreover, a person who has developed virtues will naturally act in ways consistent with moral principles. The virtuous person is the ethical person.

Civic virtues are the moral duties and rights, as a citizen of the village or the country or an integral part of the society and environment. An individual may exhibit civic virtues by voting, volunteering and organizing welfare groups and meetings. Different duties of civic virtues are:

1. To pay taxes to the local government and state, in time.
2. To keep the surroundings clean and green.
3. Not to pollute the water, land, and air by following hygiene and proper garbage disposal.
4. For example, not burning wood, tires, plastic materials, spit in the open, even not smoking in the open and not causing a nuisance to the public, are some of the civic (duties) virtues.
5. To follow the road safety rules.

There are five different rights in civic virtues and those are summarized below:

1. To vote for the local or state government.
2. To contest in the elections to the local or state government.
3. To seek a public welfare facility, such as school, hospital or a community hall or transport or communication facility, for the residents.
4. To establish a green and safe environment, pollution-free, corruption-free, and to follow ethical principles. People are said to have the right to breathe in the fresh air, by not allowing smoking in public.
5. People have an inalienable right to accept or reject a project in their area. One has the right to seek legal remedy through a public interest petition [7].

1.14   Respect for others

On personal ground respect means remembering special days, holding doors, displaying good manners, word of your genuine concern and treating others fairly. In the case of business, customers and clients will refer one's business to their friends and family, through good old-fashioned word of mouth advertising, through testimonials from satisfied customers is a powerful marketing tool. Essential features are surrounded by the following points:

1. One should not insult people or make fun of them.
2. Listen to others when they speak.
3. Value other people's opinions.
4. Considerate of people's likes and dislikes.
5. One should not mock or tease people.
6. Refrain from talking about people behind their backs.
7. Sensitive to other people's feelings.
8. One should not pressure someone to do something he or she doesn't want to do.
9. One should try to learn something from the other person.
10. One should never stereotype people.
11. Show interest and appreciation for other people's cultures and backgrounds.
12. One should not go along with prejudices and racist attitudes.

1.15   Living peacefully

One should adopt the following means to live peacefully, in the world:

1. Order in one's life (self-regulation, discipline and duty).
2. Pure thoughts in one's soul (loving others, blessing others, friendly and not criticizing or hurting others by thought, word or deed).
3. Creativity in one's head (useful and constructive).
4. Beauty in one's heart (love, service, happiness and peace).
5. Good health/body (physical strength for service).
6. Help the needy with head, heart and hands (charity). Serving the poor is considered holier than serving God.
7. Not hurting and torturing others either physically, verbally or mentally.

The following are the factors that promote living, with internal and external peace:

1. Conducive environment (safe, ventilated, illuminated and comfortable).
2. Secured job and motivated with 'recognition and reward'.
3. Absence of threat or tension by pressure due to limitations of money or time.
4. Absence of unnecessary interference or disturbance, except as guidelines.
5. Healthy labour relations and family situations.
6. Service to the needy (physically and mentally challenged) with love and sympathy.

1.16   Caring and sharing

Caring is feeling for others. It is a process that exhibits the interest in, and support for, the welfare of others with fairness, impartiality and justice in all activities, among the employees, in the context of professional ethics. It includes showing respect to the feelings of others, and also respecting and preserving the interests of all others concerned. Caring is reflected in activities, such as friendship, membership in social clubs and professional societies, and through various transactions in the family, fraternity, community, country and international councils. Different benefits of caring are given below:

1. Dependency and interdependency of human existence.
2. Emotion as the essential human nature.
3. Prioritization of human relations.
4. Family as the unit of society.
5. Interdependency of a person.

Primarily, caring influences 'sharing'. Sharing is a process that describes the transfer of knowledge (teaching, learning and information), experience (training), commodities (material possession) and facilities with others. The transfer should be genuine, legal, positive, voluntary and without any expectation in return. However, proprietary information should not be shared with outsiders. Through this process of sharing, experience, expertise, wisdom and other benefits reach more people faster. Sharing is voluntary and it cannot be driven by force, but motivated successfully through ethical principles. In short, sharing is 'charity' for humanity, 'sharing' is a culture. The 'happiness and wealth' are multiplied and the 'crimes and sufferings' are reduced, by sharing. It paves the way for peace and obviates militancy. Philosophically, sharing maximizes happiness for all human beings. In terms of psychology, the fear, divide and distrust between the 'haves' and 'have-nots' disappear. Sharing not only paves the way to prosperity, early and easily, and sustains it. Economically speaking, benefits are maximized as there is no wastage or loss, and everybody gets one's needs fulfilled and satisfied. Commercially speaking, the profit is maximized. Technologically, productivity and utilization are maximized by sharing. In the industrial arena, code-sharing in airlines for bookings on air travels and the common effluent treatment plant constructed for small-scale industries in the industrial estates, are some of the examples of sharing. The co-operative societies for producers and consumers are typical examples of sharing of the goods, profit and other social benefits. Sharing is the joint use of a resource or space. It is also the process of dividing and distributing. In our society, the protocols of communication are not based on the sharing of culture but the culture of sharing.

1.17   Self-confidence

The synergistic effect of values and goals in one's capabilities creates self-confidence. These people are usually positive thinking, flexible and willing to change. They respect others so much as they respect themselves. Self-confidence is a positive attitude, wherein the individual has some positive and realistic view of himself, to the situations in which one gets involved. The people with self-confidence exhibit courage to get into action and unshakable faith in their abilities, whatever may be their positions. They are not influenced by threats or challenges and are prepared to face them and the natural or unexpected consequences. The self-confidence in a person develops a sense of partnership, respect and accountability, and this helps the organization to obtain maximum ideas, efforts and guidelines from its employees. People with self-confidence have the following characteristics:

1. A self-assured standing.
2. Willing to listen to learn from others and adopt (flexibility).
3. Frank to speak the truth,
4. Respect others' efforts and give due credit.

On the contrary, some leaders expose others when a failure occurs, and own the credit when success comes. The factors that shape self-confidence in a person are:

1. Heredity (attitudes of parents) and family environment (elders).
2. Friendship (influence of friends/colleagues).
3. Influence of superiors/role models.
4. Training in the organization (e.g., training by Technical Evangelists at Infosys Technologies).

The following methodologies are effective in developing self-confidence in a person:

1. Encouraging SWOT analysis. By evaluating their strength and weakness, they can anticipate and be prepared to face the results.
2. Training to evaluate risks and face them (self-acceptance).
3. Self-talkis conditioning the mind for preparing the self to act, without any doubt on his capabilities. This makes one accepts himself while still striving for improvement.
4. Study and group discussion, on the history of leaders and innovators (e.g., Sam Walton of Wal-Mart, USA).

1.18   Courage

Courage is the tendency to accept and face risks and difficult tasks in rational ways. Self-confidence is the basic requirement to nurture courage. Courage is classified into three types, based on the types of risks, namely (a) physical courage, (b) social courage and (c) intellectual courage.

In physical courage, the thrust is on the adequacy of the physical strength, including the muscle power and armaments. People with high adrenalin may be prepared to face challenges for the mere 'thrill' or driven by a decision to 'excel'.

Social courage involves the decisions and actions to change the order, based on the conviction for or against certain social behaviours. This requires leadership abilities, including empathy and sacrifice, to mobilize and motivate the followers, for the social cause.

Intellectual courage is inculcated in people through acquired knowledge, experience, games, tactics, education and training. In professional ethics, courage applies to employers, employees, public and press.

One should perform Strengths, Weakness, Opportunities and Threat (SWOT) analysis. Calculate (estimate) the risks, compared with one's strengths, and anticipate the results, while taking decisions and before getting into action. Learning from the past helps. Past experience (one's own or borrowed!) and wisdom gained from self-study or others will prepare one to plan and act with self-confidence, succeed in achieving the desired ethical goals through ethical means. Opportunities and threats existing and likely to exist in the future are also to be studied and measures to be planned. This anticipatory management will help anyone to face the future with courage.

Facing criticism, owning responsibility, and accepting the mistakes or errors when committed and exposed are the expressions of courage. This sets their mind to be vigilant against past mistakes, and creative in finding the alternate means to achieve the desired objectives.

The courageous people own and have shown the following characteristics, in their professions:

(a) Perseverance (sustained hard work).

(b) Experimentation (preparedness to face the challenges, that is, unexpected or unintended results).

(c) Involvement (attitude, clear and firm resolve to act).

(d) Commitment (willing to get into action and to reach the desired goals by any alternative but ethical means).

There are 10 traits of courageous leaders as tabulated below:

1. Confront reality head-on.
2. Seek feedback and listen.
3. Say what needs to be said.
4. Encourage push-back.
5. Take action on performance issues.
6. Communicate openly and frequently.
7. Lead change.
8. Make decisions and move forward.
9. Give credit to others.
10. Hold people (and yourself) accountable.

1.19   Co-operation

It is a team spirit present with every individual engaged in engineering. Co-operation is an activity between two persons or sectors that aims at the integration of operations (synergy), while not sacrificing the autonomy of either party. Further, working together ensures, coherence, that is, blending of different skills required, towards common goals. Willingness to understand others, think and act together, and putting this into practice, is co-operation. Co-operation promotes collinearity, coherence (blend), coordination (activities linked in sequence or priority), and synergy (maximizing the output, by reinforcement). The whole is more than the sum of the individuals. It helps in minimizing the input resources (including time) and maximizes the outputs, which include quantity, quality, effectiveness and efficiency. According to professional ethics, co-operation should exist or be developed, and maintained, at several levels; between the employers and employees, between the superiors and subordinates, among the colleagues, between the producers and the suppliers (spare parts), and between the organization and its customers. The codes of ethics of various professional societies insist on appropriate co-operation to nourish the industry.

The absence of co-operation leads to lack of communication, misinformation, void in communication, and undue delay between supply, production, marketing, and consumption. This is likely to demoralize and frustrate the employees, leading to the collapse of the industry over time and an economic loss to society.

The impediments to successful co-operation are:

1. Clash of the ego of individuals.
2. Lack of leadership and motivation.
3. Conflicts of interests, based on region, religion, language and caste.
4. Ignorance and lack of interest. By careful planning, motivation, leadership, fostering and rewarding teamwork, professionalism and humanism beyond the 'divides', training on appreciation to different cultures, mutual understanding 'co-operation' can be developed and also sustained.

1.20   Commitment

Commitment means alignment to goals and adherence to ethical principles during the activities. First of all, one must believe in one's action performed and the expected results (confidence). It means one should have the conviction without an iota of doubt that one will succeed. Holding sustained interest and firmness, in whatever ethical means one follows, with the fervent attitude and hope that one will achieve the goals, is commitment. It is the driving force to realize success. This is a basic requirement for any profession. For example, a design engineer shall exhibit a sense of commitment, to make his product or project designed a beneficial contribution to society. Only when the teacher

(Guru) is committed to his job, the students will succeed in life and contribute 'good' to society. The commitment of top management will naturally lead to committed employees, whatever may be their position or emoluments. This is bound to add wealth to oneself, one's employer, society and the nation at large.

1.21   Empathy

Empathy is social radar. Sensing what others feel about, without their open talk, is the essence of empathy. Empathy begins with showing concern and then obtaining and understanding the feelings of others, from others' points of view. It is also defined as the ability to put one's self into the psychological frame of reference or point of view of another, to know what the other person feels. It includes the imaginative projection into other's feelings and understanding of other's background such as parentage, physical and mental state, economic situation and association. This is an essential ingredient for good human relations and transactions. The benefits of empathy include:

1.   Good customer relations (in sales and service, in partnering).
2.   Harmonious labour relations (in manufacturing).
3.   Good vendor–producer relationship (in partnering.)

Through the above three, we can maximize the output and profit and minimizing the loss. While dealing with customer complaints, empathy is very effective in realizing the unbiased views of others and in admitting one's limitations and failures. According to Peter Drucker, the purpose of the business is not to make a sale, but to make and keep a customer. Empathy assists one in developing courage leading to success.

1.22   Spirituality in business

Spirituality is a way of living that emphasizes the constant awareness and recognition of the spiritual dimension (mind and its development) of nature and people, with a dynamic balance between material development and spiritual development. This is said to be the great virtue of Indian philosophy and for Indians. Sometimes, spirituality includes the faith or belief in supernatural power/God, regarding worldly events. It functions as a fertilizer for the soil 'character' to blossom into values and morals. Spirituality includes creativity, communication, recognition of the individual as a human being (as opposed to a life-less machine), respect to others, acceptance (stop finding faults with colleagues and accept them the way they are), vision (looking beyond the obvious and not believing anyone blindly) and partnership (not being too authoritative, and always sharing responsibility with others, for better returns).

Spirituality is motivation as it encourages colleagues to perform better. Remember, lack of motivation leads to isolation. Spirituality is also energy: Be energetic and flexible to adapt to challenging and changing situations. Spirituality is flexible as well. One should not be too dominating. Make space for everyone and learn to recognize and accept people the way they are. Variety is the order of the day. But one can influence their mind to think and act together. Spirituality is also fun. Working is okay, but you also need to have fun in the office to keep yourself charged up. Tolerance and empathy arethe reflections of spirituality. Blue and saffron colours are said to be associated with spirituality.

Creativity in spirituality means conscious efforts to see things differently, to break out of habits and outdated beliefs to find new ways of thinking, doing and being. Suppression of creativity leads to violence. People are naturally creative. When they are forced to crush their creativity, their energy turns to destructive release and actions. Creativity includes the use of colour, humour and freedom to enhance productivity. Creativity is fun. When people enjoy what they do, it is involved. They work much harder in such a situation.

Building spirituality in the workplace is encouraged: Spirituality is promoted in the workplace by adhering to the following activities:

1. Verbally respect the individuals as humans and recognize their values in all decisions and actions.
2. Get to know the people with whom you work and know what is important to them. Know their goals, desires and dreams too.
3. State your personal ethics and beliefs clearly.
4. Support causes outside the business.
5. Encourage leaders to use value-based discretion in making decisions.
6. Demonstrate your self-knowledge and spirituality in all your actions.
7. Do unto others as you would have them do unto you.

1.23   Service learning

Service learning refers to learning the service policies, procedures, norms and conditions, other than 'the technical trade practices'. Service learning includes the characteristics of the work, basic requirements, security of the job, and awareness of the procedures while taking decisions and actions. It helps the individuals to interact ethically with colleagues, effectively coordinate with other departments, interact cordially with suppliers as well as the customers, and maintain all these friendly interactions.

Service learning may be defined as the non-paid activity, in which service is provided voluntarily to the public (have-nots in the community), non-profitable institutions and charitable organizations. It is the service during learning. This includes training or study on real-life problems and their possible solutions, during the formal learning, that is, courses of study. In the industrial scenario, adoption, study, and development of public health or welfare or safety system of a village or school is an example of service learning by the employees.

The engineering student analysing and executing a socially relevant project is another example of service learning. Service learning is a methodology falling under the category of experiential education[10,11.] It is one of the forms of experiential learning and community service opportunities. Goals met by service learning programs are:

- Intellectual development (e.g., problem solving).
- Basic skills acquisition (e.g., communication).
- Moral and ethical development.
- Social and civic responsibility.
- Career preparation.
- Multicultural understanding.
- Personal growth.

References

1. Anon, Integrity: Doing the Right Thing for the Right Reason (Canada: McGill-Queen's University Press. 2010),12. ISBN 9780773582804. [Retrieved October 15, 2013].

2. Schwartz SH (1992).'Universals in the content and structure of values: Theory and empirical tests in 20 countries', *Advances in experimental social psychology* 25, (1992):1–65.

3. Schwartz SH (2006), 'Basic human values: Theory, measurement, and applications'. *Revue Française de Sociologie*, 47: 249–288.

4. Anon, 'Human values in Ethics', Available from: https://www.gktoday.in/gk/human-values-in-ethics/ [Accessed October 10, 2019].

5. Anon, 'Ethics vs. Morals', Available from: https://www.diffen.com/difference/Ethics_vs_Morals [Accessed May 8, 2018].

6. Siva Kumar N, Rao US (1996), 'Guidelines for Value Based Management in Kautilya's Arthashastra'. *Journal of Business Ethics*, 15, 4: 415–423.

7. Nagarazan RS, *A Text Book on Professional Ethics and Human Values*, (India: Newage International (P) Ltd, India, 2006): 11, 7, 9.

8. IASPOINT, 'Integrated IAS General Studies by GK Today, August 16, 2016', Available from: https://www.gktoday.in/academy/iaspoint-details-page/[Accessed September 7, 2018].

9. 'The Five Human Values', Institute for SathyaSai Foundation, USA, Available from:http://isseusa.org/the-5-human-values [Accessed August 7, 2018].

10. Blissenden M (2006), 'Service Learning: An Example of Experiential Education in the Area of Taxation Law', *Legal Education Review*, 16, 1: Article 10, 183–193.

11. Yamuna K, 'Lecture Notes on Human Values and Professional Ethics', Department of M.B.A., Chadalawada Ramanamma Engineering College, Tirupati (A.P), 517506, India.

Engineering ethics

Abstract: Engineering ethics deals with (a) understanding the moral values that ought to guide the engineering profession, (b) resolve the moral issues in the profession and (c) justify the moral judgment concerning the profession. It is intended to develop a set of beliefs, attitudes and habits that engineers should display concerning morality. Professional societies promulgate the codes of conduct to regulate the professionals against their abuse or any unethical decisions and actions (impartiality, responsibility) affecting the individuals or groups or the society. While exercising duties, engineers often face ethical dilemmas arising out of various conflicting situations. The chapter first discusses steps involved in arriving at a reasonable judgment in considering all important factors. Engineers have to think rationally about ethical issues based on moral concerns. Since moral autonomy is at the focal point of engineering ethics, the chapter then discusses how engineers shall improve such skills to provide better solutions while resolving moral problems. Finally, different stages of moral development as per Kohlberg's and Gilligan's theory have been discussed in this chapter.

Keywords: ethical dilemma; moral autonomy; Kohlberg's theory; Gilligan's theory; consensus and controversy

2.1   Overview of engineering ethics

Engineering ethics is an activity of understanding the moral values that ought to guide the engineering profession, resolving the moral issues in the profession and justification of the moral judgment concerning the profession. Engineering ethics is the field of applied ethics and a system of moral principles that apply to the practice of engineering. The field examines and sets the obligations by engineers to society, to their clients and the profession.

2.2   Importance

Ethics guides the professional conduct of engineers. In essence, ethical values ensure honest and open transactions in the profession, and that the professionals can work without external pressure or biasedness. Ethics also ensures that engineers are held accountable for their actions, so they recognize and accept the personal commitment towards the client and the job, and maintain discretion over the client information. The role of ethics, therefore, in engineering is imperative because the integrity of the profession depends on it. The idea of not incorporating ethics into engineering is too big a risk. Only ethical engineering practices can actualize the true essence of engineering-transforming the lives of people for the greater good that have been enduring unjust hardship and difficult living conditions.

2.3   Scope

The scope of engineering ethics is two-fold:
1. Ethics of the workplace involves the co-workers and employees in an organization.
2. Ethics related to the product or work which involves the transportation, warehousing, and use, besides the safety of the end product and the environment outside the factory.

2.4   Variety of moral issues

It would be relevant to know why and how do moral issues (problems) arise in a profession or why do people behave unethically? The reasons for people including the employer and employees, behaving unethically may be classified into three categories:

1. Resource crunch

Due to pressure, through time limits, availability of money or budgetary constraints, and technology decay or obsolescence. Pressure from the government to complete the project in time (e.g., before the elections), reduction in the budget because of sudden war or natural calamity (e.g., Tsunami) and obsolescence due to technology innovation by the competitor leads to manipulation and unsafe and unethical execution of projects. Involving individuals in the development of goals and values and developing policies that allow for individual diversity, dissent and input to decision-making will prevent unethical results.

2. Opportunity

(a) Double standards or behaviour of the employers towards the employees and the public. The unethical behaviours of World Com (in the USA), Enron (in the USA as well as India) executives in 2002 resulted in bankruptcy for those companies,

(b) Management projecting their interests more than that of their employees. Some organizations over-emphasize short-term gains and results at the expense of themselves and others,

(c) Emphasis on results and gains at the expense of the employees, and

(d) Management by objectives, without focus on empowerment and improvement of the infrastructure. This is best encountered by developing policies that allow 'conscience keepers' and whistle-blowers and appointing an ombudsman, who can work confidentially with people to solve the unethical problems internally.

3. Attitude

Poor attitude of the employees set in due to

(a) low morale of the employees because of dissatisfaction and downsizing,

(b) absence of grievance redressal mechanism,

(c) lack of promotion or career development policies or denied promotions,

(d) lack of transparency,

(e) absence of recognition and reward system, and

(f) poor working environments.

Giving ethics training for all, recognizing ethical conduct in the workplace, including ethics in performance appraisal, and encouraging open discussion on ethical issues, are some of the directions to promote positive attitudes among the employees. To get a firm and positive effect, ethical standards must be set and adopted by the senior management, with input from all personnel. There are five ways to encourage a positive attitude in the workplace [1].

1. Get into the right mind set to be positive

It is a good idea for everyone to start each day with the right attitude. We often say someone 'got out of bed the wrong side in the morning' when they seem negative and critical. Although a given event or occurrence is in and of itself neutral, most people will see it as either good or bad, positive or negative depending on their mind set. If we look at events in a positive way, we are more able to accept that things happen and so we do not blame ourselves or others for it. If we can develop a healthy attitude to 'controlling the controllable' during our working day, and in life in general, we are likely to be more resilient if difficult things happen to us. We will not focus on blaming ourselves or others for these tricky situations but take a more balanced perspective. It also helps if we can understand and develop those factors on which our resilience is based such as self-belief, being able to seek support, adaptability and emotion regulation. A sensible diet, enough sleep and positive lifestyle habits such as not indulging in too much caffeine, alcohol and so on is also sensible.

2.   Support employees in becoming more positive

Helping an employee to develop their resilience levels is fundamental when it comes to managing the natural anxiety and uncertainty that are generated by certain events. This includes accepting that success will not be possible all of the time. When staff feels that negative thoughts are building up as a result of disappointment or frustration, they need to realise that they are in a bad emotional place and understand that it will not help them to resolve the problem. But they also require some form of coping mechanism. While this might involve seeking support from others, there are also opportunities for individuals to change the way that they think about problems and obstacles to approaching them from a more positive perspective. It is likewise a good idea to encourage employees to exercise or use relaxation techniques such as simple breathing exercises to help them remain calm and composed when facing difficult challenges.

3. Introduce a mentoring scheme to help staff behave more positively

When individuals are engaged in activities that they enjoy and can focus their skills and energies effectively, they tend to be more successful, have lower incidences of stress and experience a greater sense of well-being. However, there are a lot of pressures at work to distract staff from playing to their strengths. Discipline and assertiveness are needed to do so and to encourage them to enlist the help of others in areas where they are less strong. As a result, it can prove beneficial for individuals to work with other team members in reflecting on and evaluating their strengths. To help with this, you could introduce a mentoring scheme. After all, no effective sporting star or team would be without a coach today.

4. Encourage workers to record their achievements

It's important to introduce practical, achievable steps that don't take too much time to help improve employees' personal and/or professional development. This ranges from assisting them in gaining qualifications to assuming a new role or finding a better work-life balance. Having a direction and purpose helps to keep people focused when times are tough so that they can place difficulties and challenges into the broader context of achieving their goals. To this end, encourage staff to record achievements and share their goals with others to reinforce their commitment. It may be useful for them to keep a personal diary and to write in it for a few minutes each day, noting down any successes or setbacks that they have experienced. Such activity will also help them to see if they have managed to develop a more positive attitude over time.

5. Encourage employees to give something back

As human beings, we value the collaboration and support of others – no person is an island. Similarly, we also like to give something back. A lot of people who have shown single-mindedness in getting to the top of their career, vocation or sport, try to put something back in their professional area. For instance, we have seen entrepreneurs such as Bill Gates establish a global aid foundation, while the likes of Alan Sugar help young entrepreneurs to take their first step. It's important to help staff realise that, not only can they turn to others for support, but they can also help them too. Even new starters can bring fresh ideas to the team and, if employees see that they can give something back, it boosts their self-esteem.

2.5   Types of inquiry

Inquiry means an investigation. Engineering ethics involves investigations into values, meaning and facts [2]. The three types of inquiries, in solving ethical problems are: normative inquiry, conceptual inquiry and factual or descriptive inquiry.

Normative inquiry deals in identifying and justifying the morally desirable norms and standards to guide the individual or groups. These are about 'what ought to be' and 'what is good'. These questions identify and also justify the morally desirable norms or standards. Some of the questions are [3]:

(a) How far engineers are obligated to protect public safety in given situations?

(b) When should engineers start whistle-blowing on dangerous practices of their employers?

(c) Whose values are primary in taking a moral decision, employee, public or govt?

(d) Why are engineers obligated to protect public safety?

(e) When is the government justified in interfering on such issues and why?

Conceptual inquiry is meant for describing the meaning of concepts, principles, and issues related to engineering ethics. Examples are: What is safety and how is it related to risk? What does the statement, 'Protect the safety, health and welfare of the public' mean? What is a gift and what is a bribe? What is a profession and who are professionals [4]?

Factual or descriptive inquiry helps to provide facts for understanding and finding solutions to value-based issues. These are inquiries used to uncover information using scientific techniques. These inquiries get information about business realities, history of the engineering profession, procedures used in the assessment of risks and engineers psychology [3]. This may be controversial. Examples are abortion rights and the greenhouse effect.

2.6   Accepting and sharing responsibility

Responsibility is an ethical concept that refers to the fact that individuals and groups have morally based obligations and duties to others and to larger ethical and moral codes, standards and traditions. According to Aristotle, moral responsibility was viewed as originating with the moral agent as decision-maker and grew out of an ability to reason, an awareness of action and consequences and a willingness to act free from external compulsion. It is a readiness to have one's actions judged by others and, where appropriate, accept responsibility for errors, misjudgements and negligence and recognition for competence, conscientiousness, excellence and wisdom.

While working in an organization, or running a business, one needs to be a responsible player. In an organization where work quality is very important, sharing responsibility is crucial. The tasks are shared amongst team members, where it is the responsibility of every teammate to deliver flawless work from his end. Also, care needs to be taken by higher authorities that a single team player or a couple of them is not overburdened by workload.

If one of the team members is overburdened with tasks, only because having expertise in completing them flawlessly, over-work will affect productivity in the longrun. One won't be able to focus on certain tasks and apply skills due to paucity of time. Another danger of such practice is that remaining team members become complacent and lethargic. As the result, the potential of a team as a whole is reduced to a large extent. Therefore, to avoid the potential loss, each team member must be assigned a reasonable amount of workload judging his work expertise and area of specialization.

This practice of recognizing the skill set possessed by each member of the team and allocating tasks accordingly is largely the duty of the team leader. Talking again about the sense of responsibility, if every team member recognizes their responsibility towards work, the team would succeed in its endeavours. In an event of an important task coming in, tasks need to be shared amongst team members with a commitment to achieve the best results [5].

2.7   Ethical dilemmas

Dilemmas are certain kinds of situations in which a difficult choice has to be made. Moral dilemmas have two or more folding – moral obligations, duties, rights, goods or ideals come into disagreement with each other. One moral principle can have two or more conflicting applications for a particular given situation. There are three types of complexities.

Vagueness: This complexity arises due to the fact that it is not clear to individuals as to which moral considerations or principles apply to their situation (e.g.,gift – innocent amenity or unacceptable bribe).

Conflicting reasons: Even when it is clear as to which moral principle applies to one's situation, there could develop a situation wherein two or more applicable moral principles come into conflict (e.g., promise).

Disagreement:
Individuals and groups may disagree on how to interpret, apply and balance moral reasons in particular situations.

Steps/procedures in facing ethical/moral dilemmas are discussed below:

It is important to identify the relevant moral factors and reasons: i.e.,finding solutions for (i) the conflicting responsibilities, (ii) the competing rights and (iii) the clashing ideals involved.

- Collecting and gathering all the available facts which are relevant to the moral factors while resolving.
- Ranking the moral considerations or principles based on importance as applicable to the situation.
- Considering alternative courses of action for resolving the problems and tracing the full implications of each, i.e., conducting factual inquiries.
- Having talked with colleagues, friends about the problem getting their suggestions and alternative ideas on resolving that dilemma.
- Arriving at a careful and reasonable judgment or solution by taking into consideration of all important moral factors and reasons based on the facts or truths.

2.8   Moral autonomy

It is a skill and habit of thinking rationally about ethical issues based on moral concern. One can apply this moral thinking to situations that arise in professional engineering practice. Moral autonomy is at the heart of engineering ethics. Autonomy means self-governing or self-determining, i.e., act independently. Moral autonomy is concerned with the independent attitude of a person related to ethical issues. It helps to improve the self-determination among the individuals. Autonomous individuals think for themselves and do not assume that customs are always right. They seek to reason and live by general principles. Their motivation is to do what is morally reasonable for its own sake, maintaining integrity, self-respect and respect for others.

The foundation of moral autonomy is from childhood and neglected children result in the sociopath. A sociopath is the lack of sense of moral concern and guilt. Certain skills are required to improve moral autonomy. The engineers must have the ability to distinguish and relate these moral problems with the problems of law, economics, religious principles, etc. They must possess the skills of understanding,clarifying, and assessing the arguments which are against moral issues. They must have the ability to suggest the solutions to moral issues, based on facts. These suggestions must be consistent and must include all the aspects of the problem. They must have the imaginative skill to view the problems from all viewpoints and also be able to suggest a proper alternative solution. They must be able to tolerate while giving moral judgments and decisions which may cause trouble (i.e., they have to understand the difficulties in making moral decisions). They must have adequate knowledge and understanding about the use of ethical language to defend or support their views with others. They must have some better knowledge in understanding the importance of suggestions and better solutions while resolving moral problems and also about the importance of tolerance in some critical situations. They must understand the importance of maintaining moral honesty,that is, the personal convictions and beliefs and the individual's professional life must be integrated. They must have this skill of doing so.

2.9   Kohlberg's theory [6]

Moral autonomy is based on the psychology of moral development. The first psychological theory was developed by Jean Piaget based on Piaget's theory, Lawrence Kohlberg developed three main levels of moral development which are based on the kinds of reasoning and motivation adopted by individuals with regard to moral questions. Lawrence Kohlberg was, for many years, a professor at Harvard University. He became famous for his work there beginning in the early 1970s. He started as a developmental psychologist and then moved to the field of moral education. He was particularly well-known for his theory of moral development which he popularized through research studies conducted at Harvard's Center for Moral Education.

This theory is a stage theory. In other words, everyone goes through the stages sequentially without skipping any stage. However, movement through these stages is not natural, that is people do not automatically move from one stage to the next as they mature. In stage development, movement occurs when a person notices inadequacies in his or her present way of coping with a given moral dilemma.

According to stage theory, people cannot understand moral reasoning more than one stage ahead of their own. For example, a person in stage 1 can understand stage 2 reasoning but nothing beyond that. The stages include growth from self-centeredness to other-centeredness. Kohlberg proposed that moral reasoning, which he thought to be the basis for ethical behaviour, develops through stages.

- Level 1: pre-conventional
- Level 2: conventional
- Level 3: post-conventional

Stages of moral development according to Lawrence Kohlbergare shown in the below chart.

Level	Stage	Age (years)	Social orientation
Pre-conventional	1	2–4	Obedience and punishment
	2	4–7	Individualism and instrumentalism
Conventional	3	7–10	Good boy/girl
	4	10–12	Law and order
Post-conventional	5	Teens	Social contract

2.9.1   Pre-conventional level

It is nothing but a self-centered attitude. At this level, right conduct is very essential for an individual who directly benefits him. According to this level, individuals are motivated by their willingness to avoid punishment, or by their desire to satisfy their own needs. This level is related to the moral development of children and some adults who never want to go beyond a certain limit.

2.9.2   Conventional level

As per this level, the rules and norms of one's family or group or society have been accepted as the final standard of morality. When individuals are under this level, always want to satisfy others and also to meet the expectations of society and not their self-interest. No adult tries to go beyond this level.

2.9.3   Post-conventional level

This level is said to be attained when an individual recognizes the right and the wrong based on a set of principles that governing rights and the general good which are not based on self-interest or social conventions. These individuals are called 'autonomous' because they only think for themselves and also they do not agree that customs are always correct. They want to live by general principles that are universally applied to all people.

Kohlberg's theory of moral development is very much related to the goals of studying ethics at the college level Moral responsibility comes out of the foundation of early moral training given by an individual's parents and culture. As per Kohlberg's view, only a few people would reach the post-conventional level. Each level is based on the degree to which a person conforms to the conventional standards of society. Each level has two stages that represent different degrees of sophistication in moral reasoning

Research has not supported Kohlberg's belief that the development of abstract thinking in adolescence invariably leads people to the formation of idealistic moral principles. Some cross-cultural psychologists argue that Kohlberg's stories and scoring system reflect a Western emphasis on individual rights, harm and justice that is not shared in many cultures. Kohlberg's early research was conducted entirely with male subjects, yet it became the basis for a theory applied to both males and females.

2.10   Gilligan's theory [7]

2.10.1   Pre-conventional level

This is the same as Kohlberg's first level in that the person is preoccupied with self-centered reasoning, caring for the needs and desires of self.

2.10.2   Conventional

Here the thinking is opposite in that, one is preoccupied with not hurting others and a willingness to sacrifice one's interests to help or nurture others (or retain friendship).

2.10.3   Post-conventional level

This level is achieved through context-oriented reasoning, rather than by applying abstract rules ranked in a hierarchy of importance. Here the individual becomes able to strike a reasoned balance between caring about other people and pursuing one's self-interest while exercising one's rights.

2.11   Gilligan's view of Kohlberg

Carol Gilligan found that Kohlberg's theory had a strong male bias. According to Gilligan's studies, men tended to solve problems by applying abstract moral principles. Men were found to resolve moral dilemmas by choosing the most important moral rule, overriding other rules. In contrast, women gave importance to preserve personal relationships with all the people involved. The context-oriented emphasis on maintaining personal relationships was called the ethics of care, in contrast with the ethics of rules and rights adopted by men.

- Justice orientation/perspective
  Draws attention to problems of inequality and oppression and holds up an ideal of reciprocity and equal respect.
- Care orientation/perspective
  Draws attention to problems of detachment or abandonment and holds up an ideal of attention and response to need.

Gilligan states that 'Two moral injunctions – not to treat others unfairly and not to turn away from someone in need – capture these different concerns'. Gilligan's stages of development (relating to the ethics of care) is highlighted below:

Pre-conventional – striving for individual survival.

Conventional – good things come out of self-sacrifice.

Post-conventional – the principle of nonviolence toward oneself and others.

On one hand, there is a transition from selfishness to responsibility to others and on the other hand, there is the transition from goodness to truth.

2.12   Consensus and controversy

The consensus and the controversies are playing vital roles while considering moral autonomy. When an individual exercises moral autonomy, he cannot get the same results as others get in applying moral autonomy. Surely there must be some moral differences,that is, the results or verdicts will be of controversy. This kind of disagreement is unavoidable. These disagreements require some tolerances among individuals those who are autonomous, reasonable and responsible.

Consensus may be defined professionally as an acceptable resolution, one that can be supported, even if not the 'favourite' of each individual. Consensus decision-making is a group decision-making process in which group members develop, and agree to support, a decision in the best interest of the whole.

Objectives:

Agreement: seeking: a consensus decision-making process attempts to generate as much agreement as possible.

Collaborative: participants contribute to a shared proposal and shape it into a decision that meets the concerns of all group members as much as possible.

Cooperative: participants in an effective consensus process should strive to reach the best possible decision for the group and all of its members, rather than competing for personal preferences.

Egalitarian: all members of a consensus decision-making body should be afforded, as much as possible, equal input into the process. All members have the opportunity to present and amend proposals.

Inclusive: as many stakeholders as possible should be involved in the consensus decision-making process.

Participatory: the consensus process should actively solicit the input and participation of all decision-makers.

Controversy is a state of prolonged public dispute or debate, usually concerning a matter of conflicting opinion or point of view. The most applicable or well-known controversial subjects, topics or areas are politics, religion, philosophy, parenting and sex.

References

1. Anon, 'In a Nutshell: Five ways to encourage a positive attitude', Available from: https://www.hrzone.com/talent/development/in-a-nutshell-five-ways-to-encourage-a-positive-attitude [Accessed August 8, 2018].

2. Anon, Engineering Ethics – Moral Issues, Availablefrom: https://www.tutorialspoint. com/engineering_ethics/engineering_ethics_moral_issues.htm [Accessed September 12, 2018].

3. Anon, GE 20121 – Professional Ethics and Human Values, Unit 1 – Human Values, Available from:

    https://www.vidyarthiplus.com/vp/attachment.php?aid=4330[Accessed   September 20, 2018].

4. Dhakal Prabhat,'Unit – 1 Engineering Ethics', Available from:

    https://www.academia.edu/4397021/UNIT_I_Engineering_ethics_syllabus_ senses_of_engineering_ethics_variety_of_moral_issues_types_of_inquiry_moral_ dilemmas_moral [Accessed August 21, 2018].

5. Sutar Rahul, 'Sharing responsibilities at workplace', Available from: https:// ezinearticles.com/?Sharing Responsibilities-at-Workplace&id=2756140[Accessed August 8, 2018].

6. Robert N. Barger,'A summary of Lawrence Kohlberg's stages of moral development', University of Notre Dame, Notre Dame, Available from: https://www5.csudh.edu/ dearhabermas/kohlberg01bk.htm [Accessed August 8, 2018].

7. Anon, 'Gilligan's In a Different Voice', Available from: http://lfkkb.tripod.com/ eng24/gilliganstheory.html [Accessed August 8, 2018].

3

Professionalism

Abstract: Professionalism requires certain attitudes and typical qualities that are expected from a professional. Professions have collective responsibilities to promote responsible conduct by their members in many ways. One important way is to disseminate and receive seriously a code of ethics for members of the profession. This involves developing procedures for disciplining irresponsible engineers, but the main emphasis in ethics should be supporting responsible individuals. The vast majority of engineers are morally committed, but they need the support of morally committed professional societies. Professions and professionals also require thinking in terms of ethical manifestation and action aimed at preventing moral harm and unnecessary ethical problems. The chapter first discusses different criteria and characteristics of profession and professionalism. Right action theories that steer professionals in the desired direction have been classified. The chapter then discusses the code of ethics with justification, limitations and abuse. The sample code of ethics of various professional societies has also been illustrated in this chapter.

Keywords: professionalism;golden mean ethics;utilitarian theory, NSPE, IEEE, institution of engineers

3.1 Profession and professionalism

3.1.1 Profession

Profession means a job or an occupation that helps a person earn his living. The main criteria of a profession involve the following [1].

Advanced expertise − the criteria of a profession is to have sound knowledge in both technical aspects and liberal arts as well. In general, continuing education and updating knowledge are also important.

Self-regulation − an organization that provides a profession, plays a major role in setting standards for admission to the profession, drafting codes of ethics, enforcing the standards of conduct and representing the profession before the public and the government.

Public good − any occupation serves some public good by maintaining high ethical standards throughout a profession. This is a part of professional ethics where each occupation is intended to serve the welfare of the public, directly or indirectly to a possible extent.

Profession meets the criterion such as knowledge, organization and public good. In the context of knowledge, skills can't be mechanised. Examples of organizations are IEEE and ASME. Public good refers to protect public health safety welfare.

3.1.2   Professional

A person who is paid for getting involved in a particular profession to earn a living as well as to satisfy the laws of that profession can be regarded as a professional. The definition of a professional is given differently by different experts in the field. Let us observe the following definitions [1]:

'Only consulting engineers who are basically independent and have freedom from coercion can be called as professionals.' – **Robert L. Whitelaw**

'Professionals have to meet the expectations of clients and employers. Professional restrains are to be imposed by only laws and government regulations and not by personal conscience.' – **Samuel Florman**

'Engineers are professionals when they attain standards of achievement in education, job performance or creativity in engineering and accept the most basic moral responsibilities to the public as well as employers, clients, colleagues and subordinates.'–**Mike martin and Ronald Schinzinger**

3.1.3   Professionalism

The criteria for achieving and sustaining professional status or professionalism are [2]:

1. Advanced expertise: The expertise includes sophisticated skills and theoretical knowledge in exercising judgment. This means that a professional should analyse the problem in a specific known area in an objective manner.

2. Self-regulation: One should analyse the problem independently of self-interest and direct to a decision towards the best interest of the clients/customers. An autonomous judgment (unbiased and on merits only) is anticipated. In such situations, the codes of conduct of professional societies are followed as guidance.

3. Public good: One should not be a mere paid employee of an individual or a teaching college or manufacturing organization, to execute whatever the employer wants one to do. The job should be recognised by the public. The concerted efforts in the job should be towards the promotion of the welfare, safety and health of the public.

3.1.3.1   Characteristics

The characteristics of the 'profession' as distinct from 'non-professional occupation' are listed as follows [2]:

1. Extensive training

Entry into the profession requires an extensive period of training of intellectual (competence) and moral (integrity) character. The theoretical base is obtained through formal education, usually in an academic institution. It may be a bachelor's degree from a college or university or an advanced degree conferred by professional schools.

2. Knowledge and skills

Knowledge and skills (competence) are necessary for the well-being of society. Knowledge of physicians protects us from disease and restores health. The lawyer's knowledge is useful when we are sued for a crime, or if our business is to be merged or closed or when we buy a property. The Chartered Accountant's knowledge is important for the success of recording financial transactions or when we file the income return. The knowledge, study and research of the engineers are required for the safety of the airplane, for technological advances and national defence.

3. Monopoly

Monopoly control is achieved in two ways:

(a) The profession convinces the community that only those who have graduated from the professional school should be allowed to hold the professional title. The profession also gains control over professional schools by establishing accreditation standards.

(b) By persuading the community to have a licensing system for those who want to enter the profession. If practicing without a license, they are liable to pay penalties.

4. Autonomy in the workplace

Professionals engaged in private practice have considerable freedom in choosing their clients or patients. Even the professionals working in large organizations exercise a large degree of impartiality, creativity and discretion (care with decision and communication) in performing their responsibilities. Besides this, professionals are empowered with certain rights to establish their autonomy. Accordingly, physicians must determine the most appropriate medical treatments for their patients and lawyers must decide on the most successful defence for their clients. The possession of specialised knowledge is thus a powerful defence of professional autonomy.

5. Ethical standards

Professional societies disseminate the codes of conduct to regulate the professionals against their abuse or any unethical decisions and actions affecting the individuals or groups or the society.

3.1.4   Motives for professionalism

Motives for professionalism are a desire for interesting and challenging work. The pleasure in the act of changing the world, the joy of creative efforts where a scientist's interest is in discovering new technology and engineer's interest is derived from creatively solving practical problems. The engineer shares the scientist's job in understanding the laws and riddles of the universe. The sheer magnitude of nature viz. oceans, rivers, mountains and prairies leads engineers to build engineering marvels like ships, bridges, tunnels, etc., which appeal to human passion.

3.1.5   Models of professional roles

The promotion of public good is the most important concern of professional engineers. There are several role models to whom the engineers are attracted. These models provoke their thinking, attitudes and actions [1].

1. Saviour

The engineer as a saviour saves the society from poverty, illiteracy, wastage, inefficiency, ill health, human (labour) dignity and leads it to prosperity, through technological development and social planning.

2. Guardian

He guards the interests of the poor and general public as one who is conversant with technology development, is given the authority befitting his expertise to determine what is best suited to the society.

3. Bureaucratic servant

He serves the organization and the employers. The management of an enterprise fixes its goals and assigns the job of problem-solving to the engineer, who accepts the challenge and shapes them into concrete achievements.

4. Social servant

It exhibits social responsibility. The engineer translates the interest and aspirations of the society into a reality, remembering that his true master is the society at large.

5. Social enabler and catalyst

He is the one who changes society through technology. The engineer must assist the management and the society to understand their needs and make informed decisions on the desirable technological development and minimise the negative effects of technology on people and their living environment. Thus, he shines as a social enabler and a catalyst for further growth.

6. Game player

He is neither a servant nor a master. An engineer is an assertive player, not a passive player who may carry out his master's voice. He plays a unique role successfully within the organization, enjoying the excitement of the profession and having the satisfaction of surging ahead in a competitive world.

3.2   Right action theories

There are four types of theories on ethics, which help to create the fundamental principles of obligation suitable and applicable to the professional and personal conduct of a person in his everyday life. These theories are essential for the cause of right action and morality. They are as follows [2]:

- Golden mean ethics (Aristotle, 384–322 BC)
- Rights-based ethics (John Locke, 1632–1704) (a) Liberty-rights, (b) Welfare rights
- Duty-based ethics (Immanuai Kant, (1724–1804)
- Utilitarian ethics (John Stuart Mill (1806–1873) (a) Act-utilitarianism, (b) Rule-utilitarianism

3.2.1   Golden mean ethics

Golden mean ethics justify the moral obligations of an engineer. The best solution is achieved through reason and logic and is a compromise or golden mean between extremes of excess and deficiency. For example, in the case of the environment, the golden mean between the extremes of 'neglect' and 'exploitation' might be 'protection'. This emphasizes the character rather than the rights or duties. The character is the pattern of virtues (morally desirable features) and that is why it is also termed as 'virtue theory'. The theory advocated by Aristotle stressed the tendency to act at a proper balance between extremes of conduct, emotion, desire, attitudes to find the golden mean between the extremes of 'excess' or 'deficiency'. Problem encounters here are a variation from one person to another in their powers of reasoning and the difficulty in applying the theory to ethical problems.

On the other hand, the virtue theory proposed by Scottish philosopher *MacIntyre* highlighted on the actions aimed at achieving common good and social (internal) good such as social justice, promotion of health, creation of useful and safe technological products and services. Five types of virtues that constitute responsible professionalism are public-spirited virtues, proficiency virtues, teamwork virtues, self-governance virtues and cardinal virtues.

3.2.2   Right-based ethics

The rights approach to ethics has its roots in the 18th-century philosopher Immanuel Kant, who focused on the individual's right to choose for oneself. According to him, what makes human beings different from mere things is, that people have dignity based on their ability to choose freely what they will do with their lives, and they have a fundamental moral right to have these choices respected. People are not objects to be manipulated; it is a violation of human dignity to use people in ways they do not freely choose. Other rights he advocated are:

1. The right to access the truth:
We have a right to be told the truth and to be informed about matters that significantly affect our choices.

2. The right to privacy:
We have the right to do, believe and say whatever we choose in our personal lives so long as we do not violate the rights of others.

3. The right not to be injured:
We have the right not to be harmed or injured unless we freely and knowingly do something to deserve punishment or we freely and knowingly choose to risk such injuries.

4. The right to what is agreed:

We have a right to what has been promised by those with whom we have freely entered into a contract or agreement.

The rights theory as promoted by John Locke states that the actions are right if they respect the human rights of everyone affected. He proposed the three basic human rights, namely life, liberty and property. His views were reflected in modern American society when Jefferson declared basic rights as life, liberty and pursuit of happiness.

As per Philosopher A.I. Melden'stheory based on rights [3], nature mandates that we should not harm others' life, health, liberty or property. Melden allowed welfare rights also for living a decent human life. He highlighted that the rights should be based on the social welfare system.

3.2.2.1   Human rights

Human rights are explained in two forms, namely liberty rights and welfare rights. Liberty rights are rights to exercise one's liberty and stresses duties on other people not to interfere with one's freedom. The four features of liberty rights (also called moral rights), which lay the base for Government administration, are:

1. Rights are natural in so far as they are not invented or created by the government.

2. They are universal, as they do not change from country to country.

3. They are equal since the rights are the same for all people, irrespective of caste, race, creed or sex.

4. They are inalienable, i.e., one cannot hand over his rights to another person, such as selling oneself to slavery.

Welfare rights are the rights to benefit the needy for a decent human life, when one cannot earn those benefits and when those benefits are available in the society.

3.2.2.2   Economic rights

In the free-market economy, the very purpose of the existence of the manufacturer, the sellers and the service providers is to serve the consumer. The consumer is eligible to exercise some rights. The consumer's six basic rights are Right to Information, Right to Safety, Right to Choice, Right to be Heard, Right to Redressal and Right to Consumer Education.

3.2.3   Duty-based ethics

A. The duty ethics theory, proposed by German philosopher Immanuel Kant (1724–1804) [3] states that actions are consequences of performance of one's duties, such as 'being honest', 'not cause suffering of others', 'being fair to others including the humble and weak', 'being grateful', 'keeping promises', etc. The stress is on the universal principle of respect for autonomy, that is, respect and rationality of persons. As per Kant, we have duties to ourselves, as we are rational and autonomous beings. We have a duty not to commit suicide; a duty to develop our talents and a duty to avoid harmful drugs. Kant insisted that moral duties are categorical imperatives. They are commands that we impose on ourselves as well as other rational beings. For example, we should be honest because honesty is required by duty. A businessman is to be honest because honesty pays in terms of profits from customers and from avoiding jail for dishonesty.

B. On the other hand, the duty ethics theory, as enunciated by American moral and political philosopher John Rawl [3], gave importance to the actions that would be voluntarily agreed upon by all persons concerned, assuming impartiality. His view emphasized the autonomy each person exercises in forming agreements with other rational people. Rawl proposed two basic moral principles:

1. Each person is entitled to the most extensive amount of liberty compatible with an equal amount for others.
2. Differences in social power and economic benefits are justified only when they are likely to benefit everyone, including members of the most disadvantaged groups.

The first principle is of prime importance and should be satisfied first. Without basic liberties, other economic or social benefits cannot be sustained for long. The second principle insists that to allow some people with great wealth and power is justified only when all other groups are benefited. In the business scenario, for example, free enterprise is permissible so far it provides the capital needed to invest and prosper, thereby making job opportunities to the public and taxes to fund the government spending on the welfare schemes on the poor people.

C. W. D. Ross, the British philosopher [3] introduced the term prima facie duties, which means duties might have justified exceptions. Most duties are prima facie ones; some may have obligatory or permissible exceptions. Ross assumed that the prima facie duties are intuitively obvious (self-evident) while fixing priorities among duties. He noted that the principles, such as 'do not kill' and 'protect innocent life' involve high respect for persons than other principles such as, 'do not lie' (less harmful). This theory is criticised on the fact, that the intuitions do not provide sufficient guidelines for moral duty. He has listed various aspects of duty ethics that reflect our moral convictions, namely:

1. Fidelity: duty to keep promises.
2. Reparation: duty to compensate others when we harm them.
3. Gratitude: duty to thank those who help us.
4. Justice: duty to recognize merit.
5. Beneficence: duty to recognise inequality and improve the condition of others.
6. Self-improvement: duty to improve virtue and intelligence.
7. Non-malfeasance: duty not to injure others.

3.2.4   Utilitarian theory

The word utilitarianism was conceived in the 19th century by English philosopher, jurist and social reformer Jeremy Bentham and British philosopher, political economist, and civil servant John Stuart Mill [3] to help legislators determine which laws were ethically best. They suggested that the standard of right conduct is the maximization of good consequences which mean either 'utilities' or the 'balance of good over evil'. This approach relates to the costs and benefits. Right actions are the ones that produce the greatest satisfaction of the preferences of the affected persons. In analysing an issue in this approach, one has to:

(a) Identify the various courses of action available to us.

(b) Ask who will be affected by each action and what benefits or harms will be derived from each.

(c) Choose the action that will produce the greatest benefits and the least harm. The ethical action is the one that provides the greatest good for the greatest number.

The act utilitarian theory projected by British philosopher, political economist, and civil servant J.S. Mill (1806–1873) [3] focuses on actions, rather than on general rules. An action is right if it generates the most overall good for the most people involved.

The rule utilitarian theory, developed by American philosopher Richard Brandt (1910–1997) [3] stressed the rules, such as 'do not steal', 'do notharm others', 'do not bribe', as of primary importance. He suggested that individual actions are right when they are required by a set of rules which maximise the public good.

The act utilitarian theory allowed a few immoral actions. Hence, there was a need to develop rule utilitarian theory *to* institute morality and justice, For instance, stealing an old computer from the employer will benefit the employee more than the loss to the employer. This action is right as per Act Utilitarian. But Rule Utilitarian observes this as wrong because the employee should act as 'faithful agent or trustee of the employees'. In another illustration, some undisciplined engineers are terminated with the blame for the mistakes they have not committed. The process is unreasonable although this results in the promotion of overall good.

Apart from the four basic theories, there are two other theories in existence

3.2.4.1   Justice (fairness) theory [3]

The justice or fairness approach to ethics has its roots in the teachings of the ancient Greek philosopher Aristotle, who said that 'equals should be treated equally and unequal unequally.' The basic moral issue in this approach is: how fair is an action? Does it treat everyone in the same way, or does it shows favouritism and discrimination?

3.2.4.2   Self-realization ethics [3]

Right action consists of seeking self-fulfilment. In one version of this theory, the self to be realized is defined by caring relationships with other individuals and society. In another version called ethical egoism, the right action consists in always promoting what is good for oneself. No caring and society relationships are assumed.

3.3   Uses of ethical theories and their limitations

The uses of ethical theories are listed below:
  (i)   Understanding and resolving moral dilemmas.
  (ii) Estimating the professional obligations and ideals.
  (iii) Determining to what extent, the obligations can be exercised in a given situation.
  (iv) Interpreting the facts of a given situation and finding out who is right and who is wrong.
  (v)   Providing a conceptual base for evaluation and analysis of facts and circumstances.

(vi) Providing a relative ranking of various dilemmas involved in a case.

For example, if a given situation is interpreted by applying the 'Theory of virtue ethics', the following aspects will become important: loyalty to the public (vs) loyalty to the organization. By applying 'utilitarian theory,' the same issue will be examined in terms of, public good vs the economic need of the organization. Right ethics' followers will examine the same issue in terms of rights of the public vs the rights of the management. Duty ethics' followers will examine the aspects, duty of the organization to protect the public vs duty of the public to respect the organization's rights to earn its profits.

The limitations of the code of ethics are as follows:

- Codes are general and vague.
- Some codes conflict with each other.
- They cannot serve as the final moral authority for professional conduct. There is the proliferation of codes.

3.4   Senses of corporate responsibility

Corporate social responsibility (CSR) is a business approach that contributes to sustainable development by delivering economic, social and environmental benefits for all stakeholders. CSR is a perception with many definitions and practices. The way it is understood and implemented differs greatly for each company and country. Moreover, CSR is a very broad concept that addresses many and various topics, such as human rights, corporate governance, health and safety, environmental effects, working conditions and contribution to economic development. Whatever the definition is, the purpose of CSR is to drive change towards sustainability. A company›s sense of responsibility towards the community and both ecological and social environment is important in which it operates. Companies express this citizenship, (1) through their waste and pollution reduction processes, (2) by contributing educational and social programs and (3) by earning adequate returns on the employed resources [4].

3.5   Codes of ethics

Engineers, scientists and technologists influence the quality of life for all people in the complex technological society. In the quest of their profession, therefore, it is vital that members ethically conduct their work so that they value the confidence of colleagues, employers, clients and the public. Engineers must perform under a standard of professional behaviour that requires loyalty to the highest principles of ethical conduct on behalf of the public, clients, employers and the profession.

3.5.1   Importance

Engineering is an important and cultured profession. As members of this profession, engineers are expected to demonstrate the highest standards of honesty and integrity. Code of ethics defines the values and principles that shape the decisions engineers create in engineering practice. The related guidelines on professional conduct offer a framework when exercising their judgment.

3.5.2   Justification

Business organizations often develop several different policies, rules and guidelines for leading their operations. While home-based or sole proprietorship businesses usually require fewer policies, larger organizations use these guidelines to manage employee behaviour. A code of ethics is a common organizational policy used in business organizations. The code of ethics policy usually sets the minimum standards for business owners, managers and employees to follow when completing various business functions. Code of ethics can help companies improve business relationships.

3.5.3   Limitations

A code of ethics is not a legal document. One cannot be arrested for violating its provisions, although expulsion from the professional society might result from code violations. With the current state of engineering societies, expulsion from an engineering society generally will not result in an inability to practice engineering, so there are not necessarily any direct consequences of violating engineering ethical codes. Relatively few practicing engineers are members of professional societies and so donot necessarily feel compelled to abide by their codes. Many engineers who are members of professional societies are not aware of the existence of the society's code, or if they are aware of it, they have never read it. Even among engineers who know about their society's code, meeting the code is rare. There are also objections that the engineering codes often have internal conflicts but donot give a method for resolving the conflict. Finally, codes can be coercive: they cultivate ethical behaviour with a stick rather than with a carrot [5].

3.5.4   Abuse

3.5.4.1   Unenforceability

Some codes of conduct, perhaps because of employee delinquency in the past, take a micro management approach, dictating detailed minutia like the kinds of material that can be worn in the office or the exact length of personal phone calls. According to leadership skills for life, codes of conduct need to be detailed because some questions, like whether or not taking a company pen home is ethical, will deliver many answers. Codes can be cumbersome, contradictory and ultimately ineffective when people, including supervisors who cannot enforce the codes and still have a productive workplace, abandon them in favour of 'common sense'. On the contrary, a company in which the value of honesty is embedded and appreciated can result in a culture where no one would consider taking a pen.

3.5.4.2   Inequality

Codes of conduct are often framed, in part, to ensure that all members of an organization are treated equally. However, often those in upper-level management and creative positions are given a 'bye' on certain codes, like those restricting how the worker talks about the company or to what degree employees are allowed to have personal relationships outside of work.

3.5.4.3   Unethical corporate behaviour

Codes of conduct that limit employees' ability to speak out against the corporation can keep them quiet for fear of job loss or legal retribution even if the company is engaging in an unethical practice.

3.6   NSPE (National Society of Professional Engineers) code of ethics [6]

3.6.1   Preamble

Engineering is an important and learned profession. As members of this profession, engineers are expected to exhibit higher standards of honesty and integrity. Engineering has a direct and vital impact on the quality of life for all people. Accordingly, the services provided by engineers require honesty, impartiality, fairness and equity, and must be dedicated to the protection of public health, safety and welfare. The engineer must perform under a standard of professional behaviour that requires adherence to the highest principles of ethical conduct.

I. Fundamental seven canons

Engineers in the fulfilment of their professional duties shall

- hold paramount the safety, health and welfare of the public.
- perform services only in areas of their competence.
- issue public statements only in an objective and truthful manner.
- act for each employer or client as faithful agents or trustees.
- avoid deceptive acts (shall build their professional reputation on the merit of their services and shall not compete unfairly with others).
- conduct themselves honourably, responsibly, ethically and lawfully to enhance the honour, reputation and usefulness of the profession (shall act with zero tolerance for bribery, fraud and corruption).
- shall continue their professional development throughout their career and shall provide opportunities for the professional development of those engineers under their supervision.

II. Rules of practice

Engineers shall hold paramount the safety, health and welfare of the public.

- Engineers shall perform services only in the areas of their competence.
- Engineers shall undertake assignments only when qualified by education or experience in the specific technical fields involved.
- Engineers shall issue public statements only in an objective and truthful manner.
- Engineers shall be objective and truthful in professional reports, statements or testimony.
- They shall include all relevant and pertinent information in such reports, statements, or testimony, which should bear the date indicating when it was current.
- Engineers shall act for each employer or client as faithful agents or trustees.
- Engineers shall disclose all known or potential conflicts of interest that could influence or appear to influence their judgment or the quality of their services.
- Engineers shall avoid deceptive acts.
- Engineers shall not falsify their qualifications or permit misrepresentation of their or their associates' qualifications.
- They shall not misrepresent or exaggerate their responsibility in or for the subject matter of prior assignments.

- Brochures or other presentations incident to the solicitation of employment shall not misrepresent pertinent facts concerning employers, employees, associates, joint ventures or past accomplishments.

III. Professional obligations

- Engineers shall be guided in all their relations by the highest standards of honesty and integrity.
- Engineers shall at all times strive to serve the public interest.
- Engineers are encouraged to participate in civic affairs.
- Engineers shall provide career guidance for youths.
- Engineers shall work for the advancement of the safety, health and well-being of their community.
- Engineers shall avoid all conduct or practice that deceives the public.
- Engineers shall avoid the use of statements containing a material misrepresentation of fact or omitting a material fact.
- Engineers shall not disclose, without consent, confidential information concerning the business affairs or technical processes of any present or former client or employer or public body on which they serve.
- Engineers shall not, without the consent of all interested parties, promote or arrange for new employment or practice in connection with a specific project for which the engineer has gained particular and specialized knowledge.
- Engineers shall not be influenced in their professional duties by conflicting interests.
- Engineers shall not accept financial or other considerations, including free engineering designs, from material or equipment suppliers for specifying their product.
- Engineers shall not attempt to obtain employment or advancement or professional engagements by untruthfully criticizing other engineers, or by other improper or questionable methods.
- Engineers shall not request, propose, or accept a commission on a contingent basis under circumstances in which their judgment may be compromised.
- Engineers shall not attempt to injure, maliciously or falsely, directly or indirectly, the professional reputation, prospects, practice or employment of other engineers.

- Engineers who believe others are guilty of unethical or illegal practice shall present such information to the proper authority for action.
- Engineers in private practice shall not review the work of another engineer for the same client, except with the knowledge of such engineer, or unless the connection of such engineer with the work has been terminated.
- Engineers shall accept personal responsibility for their professional activities, provided, however, those engineers may seek indemnification for services arising out of their practice other than gross negligence, where the engineer's interests cannot otherwise be protected.
- Engineers shall conform to state registration laws in the practice of engineering.
- Engineers shall give credit for engineering work to those to whom credit is due, and will recognize the proprietary interests of others.
- Engineers shall, whenever possible, name the person or persons who may be individually responsible for designs, inventions, writings, or other accomplishments.

3.7   The Institute of Electrical and Electronics Engineers (IEEE) code of ethics [7]

Article I

Members shall maintain high standards of diligence, creativity and productivity, and shall:

- Accept responsibility for their actions;
- Be honest and realistic in stating claims or estimates from the available date.
- Undertake technological tasks and accept responsibility only if qualified by training or experience, or after full disclosure to their employers or clients of pertinent qualifications.
- Maintain their professional skills at the level of the state of the art, and recognize the importance of current events in their work.
- Advance the integrity and prestige of the profession by practicing in a dignified manner and for adequate compensation.

Article II

Members shall, in their work:

- Treat fairly all colleagues and co-workers, regardless of race, religion, sex, age or national origin;
- Report, publish and disseminate freely information to others, subject to legal and proprietary restraints; encourage colleagues and co-workers to act in accord with this code and support them when they do so.
- Seek, accept and offer honest criticism of work, and properly credit the contributions of others.
- Support and participate in the activities of their professional societies.
- Assist colleagues and co-workers in their professional development.

Article III

Members shall, in their relations with employers and clients:

- Act as faithful agents or trustees for their employers or clients in professional and business matters, provided such actions conform with other parts of this code;
- Keep information on the business affairs or technical processes of an employer or client in confidence while employed, and later, until such information is properly released, provided such actions conform with other parts of this code;
- Inform their employers, clients, professional societies or public agencies or private agencies of which they are members or to which they may make presentations, of any circumstance that could lead to a conflict of interest.
- Neither give nor accept, directly or indirectly, any gift, payment or service of more than nominal value to or from those having business relationships with their employers or clients.
- Assist and advise their employers or clients in anticipating the possible consequences, direct and indirect, immediate or remote, of the projects, work or plans of which they have knowledge.

Article IV

Members shall, in fulfilling their responsibilities to the community:

- Protect the safety, health and welfare of the public and speak out against abuses in these areas affecting the public interest.
- Contribute professional advice, as appropriate, to civic, charitable or other non-profit organizations.

- Seek to extend public knowledge and appreciation of the profession and its achievements.

3.8   Institution of Engineers (India) code of ethics [8]

1. The Corporate Members of The Institution of Engineers (India) are committed to promote and practice the profession of engineering for the common good of the community bearing in mind the following concerns:

    (i)   Concern for ethical standards.

    (ii) Concern for social justice, social order and human rights.

    (iii) Concern for protection of the environment.

    (iv) Concern for sustainable development.

    (v)   Public safety and tranquillity.

2.

    (i)   A corporate member shall utilise his knowledge and expertise for the welfare, health and safety of the community without any discrimination for sectional or private interests.

    (ii) A corporate member shall maintain honour, integrity and dignity in all his professional actions to be worthy of the trust of the community and the profession.

    (iii) A corporate member shall act only in the domains of his competence and with diligence, care, sincerity and honesty.

    (iv) A corporate member shall apply his knowledge and expertise in the interest of his employer or the clients for whom he shall work without compromising with other obligations to these tenets.

    (v)   A corporate member shall not falsify or misrepresent his own or his associates' qualifications, experience, etc.

    (vi) A corporate member, wherever necessary and relevant, shall take all reasonable steps to inform himself, his employer or clients, of the environmental, economic, social and other possible consequences, which may arise out of his actions.

    (vii) A corporate member shall maintain utmost honesty and fairness in making statements or giving witness and shall do so based on adequate knowledge.

    (viii) A corporate member shall not directly or indirectly injure the professional reputation of another member.

(ix) A corporate member shall reject any kind of offer that may involve an unfair practice or may cause avoidable damage to the ecosystem.

(x) A corporate member shall be concerned about and shall act to the best of his abilities for maintenance of sustainability of the process of development.

(xi) A corporate member shall not act in any manner which may injure the reputation of the Institution or which may cause any damage to the Institution financially or otherwise.

3. General guidance

The tenets of the Code of Ethics are based on the recognition that

(i) A common tie exists among humanity and that The Institution of Engineers (India) derives its value from the people so that the actions of its corporate members should indicate the member's highest regard for equality of opportunity, social justice and fairness.

(ii) The corporate members of the institution hold a privileged position in the community to make it a necessity for their not to use the position for personal and sectional interests.

4. And, as such, a corporate member

(i) should keep his employer or client fully informed on all matters in respect of his assignment which is likely to lead to a conflict of interest or when, in his judgment, a project will not be viable based on commercial, technical, environmental or any other risks.

(ii) should maintain the confidentiality of any information with utmost sincerity unless expressly permitted to disclose such information or unless such permission if withheld, may adversely affect the welfare, health and safety of the community.

(iii) should neither solicit nor accept financial or other considerations from anyone related to a project or assignment of which he is in charge.

(iv) should neither pay nor offer direct or indirect inducements to secure work.

(v) should compete based on merit alone.

(vi)   should refrain from inducing a client to breach a contract entered into with another duly appointed engineer.

(vii)   should if asked by the employer or a client, review the work of another person or organization, discuss the review with the other person or organization to arrive at a balanced opinion.

(viii) should make statements or give evidence before a tribunal or a court of law in an objective and accurate manner and express any opinion based on adequate knowledge and competence.

(ix)   should reveal the existence of any interest – pecuniary or otherwise – which may affect the judgment while giving evidence or making a statement.

5. Any decision of the council as per provisions of the relevant bye-laws of the institution shall be final and binding on all corporate members.

References

1. Tutorials point: Simply easy learning, 'Professions and professionalism', Available from:

    https://www.tutorialspoint.com/engineering_ethics/professions_and_professionalism.htm [Accessed August 19, 2018].

2. Naagarazan R.S., *Professional ethics and human values* (New Delhi: Newage International Publishers, 2006), 30–40.

3. Anon, Theories of about right action (Ethical Theories), Available from: https://www.snscourseware.org/snsce/files/CW_5864b96198d81/theories%20about%20right%20action.pdf [Accessed September 8, 2018].

4. Business dictionary, corporate social responsibility, Available from:

    http://www.businessdictionary.com/definition/corporate-social-responsibility.html [Accessed August 11, 2018].

5. Charles E. Harris, Michael S. Pritchard, Michael J. Rabins, *Engineering ethics: concepts and cases* (Australia: National Library, 1995), 0–411.

6. National Society of Professional Engineers, 'NSPE code of ethics for engineers', Available from: https://www.nspe.org/resources/ethics/code-ethics [Accessed August 11, 2018].

7. The Institute of Electrical and Electronics Engineers, 'IEEE code of ethics for engineers', Available from: http://icrom.ir/files/IEEE%20code%20of%20Ethics.pdf [Accessed August 12, 2018].

8. The Institution of Engineers in India, 'Code of ethics', Available from: https://www.ieindia.org/webui/iei-home.aspx [Accessed August 18, 2018].

4

Engineering as social experimentation

Abstract: Engineering is an inherently risky activity, keeping in view of the fact that all products of technology present some latent dangers. To accentuate this fact and help in exploring its ethical implications, it is recommended that engineering should be viewed as an investigational process. It is not, of course, an experiment conducted solely in a laboratory under controlled conditions. To a certain extent, it is an experiment on a social scale involving human subjects. The chapter first discusses engineering as social experimentation with conscientious, moral sovereignty and accountability. Safety is a fundamental part of engineering design. Engineers have a responsibility to society to produce reasonably safe products. Many disasters have happened in the past due to safety reasons. The chapter then reviews such types of incidents with safety lessons learned and exercise precautions and control measures by assessing risk.

Keywords: engineering as experimentation; conscientiousness; moral sovereignty; balanced outlook on law; cautious optimism; risk assessment

4.1   Engineering as experimentation

It is universally true that to undertake a great and novel work, an experiment needs to be carried out. It means taking up a struggle with the forces of nature without the assurance of emerging as the victor after the first attack. Experimentation takes an important and useful role in the design process [1]. All products of technology present some probable dangers, and thus engineering is an inherently risky activity. To underscore this fact and help in exploring its ethical implications, we suggest that engineering should be viewed as an experimental process. It is not, of course, an experiment conducted solely in a laboratory under controlled conditions. Rather, it is an experiment on a social scale involving human subjects. Engineering has a direct and vital effect on the quality of life of people. Accordingly, the services provided by engineers must be dedicated to the protection of public safety, health and welfare.

4.1.1   Difference between engineering and experimentation

The objective of engineering is to solve problems related to unknowns, uncertain outcomes, monitor, learn from past experiments, human subjects/

participants often unaware, uninformed, and often do not recognize all variables of natural experiment.

On the contrary, the objective of experimentation is to find new knowledge or answers which also involve: unknowns, uncertain outcome, test hypothesis, draw conclusions or verify hypothesis based on experience/ evidence, 'informed consent' of subjects, and try to control all variables controlled experiment.

Informed consent consists of knowledge and voluntariness. The subjects (human beings) should be supplied with all the information needed to make a reasonable decision before getting into the experiment without being subjected to force, fraud or deception. Supplying complete information is neither necessary nor in most cases possible. But all relevant information required for making a logical decision on whether to participate should be conveyed. Generally, we all prefer to be the subject of our experiments rather than those of others.

In standard experiments, members belong to two different groups. Members of one group receive special experimental treatment. The other group members called the 'control group' do not receive special treatment, though they are from the same environment in all other respects. But this is not true in engineering, since most of the experiments are not conducted in laboratories. The subjects of experiments are human beings who are outside the experimenter's control. Thus, it is not possible to study the effects of changes in the variable on different groups. Hence, only historical and retrospective data available about various target groups has to be used for evaluation. Thus engineering as social experimentation seems to be an extended usage of the concept of experimentation.

Experiments are carried out in fractional ignorance when outcomes are uncertain. The engineers are asked to make things work without all the accessible scientific knowledge, safety facts, environment, health, social influences, etc., or when good design relies on information gathered before and after a product leaves the factory. The product is tested in its true 'environment,' not counterfeit ones used to simulate the real environment.

4.1.2   Engineers as responsible experimenters [2]

Although the engineers assist in experiments, they are not alone in the field. Their duty is shared with the organizations, people, government and others. Undoubtedly the engineers share a greater responsibility while monitoring the projects, identifying the risks and informing the clients and the public with facts. Based on this, they used to make decisions to participate or protest or promote.

The engineer, as an experimenter, has several responsibilities to society, namely,

1. A conscientious commitment to live by moral values.
2. A comprehensive perspective on relevant information which includes constant awareness of the progress of the experiment and readiness to monitor the side effects, if any.
3. Unrestricted free-personal attachment in all steps of the project/ product development(autonomy).
4. Be responsible for the results of the project (accountability).

4.2   Conscientiousness

Conscientious moral obligation means: (a) being sensitive to the full range of moral values and responsibilities relevant to the prevailing situation and (b) the keenness to develop the skill and put efforts needed to reach the best balance possible among those considerations. In a nutshell, engineers must possess open eyes, open ears and an open mind (i.e., moral vision, moral listening and moral reasoning). This makes the engineers as social experimenters, respect the safety and health of the affected, while they seek to enrich their knowledge, rush for profit, follow the rules, or care for only the recipient. The human rights of the participant should be protected through voluntary and informed consent.

4.3   Broad outlook

The engineer should grasp the context of his work and ensure that the work involved results in only moral ends. One should not ignore his ethics, if the product or project that he is involved in will result in damaging the nervous system of the people (or even the enemy, in case of weapon development) ora product has a built-in outmoded component to boost sales with a false claim. In possessing the perspective of realistic information, the engineer should exhibit a moral concern and not agree with this design. Sometimes, the blame is transferred to the government or the competitors. Some organizations think that they will let the government find the fault or let the deceptive competitor be caught first. Finally, a full-scale environmental or social impact study of the product or project by individual engineers is helpful but not achievable, in practice.

4.4   Moral sovereignty

Viewing engineering as social experimentation will promote independent participation and retain one's professional identity. Periodical performance appraisals, tight-time schedules, and fear of foreign competition threaten this sovereignty. The attitude of the management should allow liberty in the judgments of their engineers on moral issues. If management views profitability as more important than reliable quality and retention of the customers that discourage moral sovereignty, engineers are compelled to seek support from their professional societies and outside organizations for moral support. Taking into consideration that engineering is social experimentation and anticipating unknown consequences, one should promote an attitude of questioning about the adequacy of the existing economic and safety standards. This proves a greater sense of personal attachment in one's work.

4.5   Accountability

The term accountability means:
1.   The capacity to understand and act on moral reasons.
2.   Willingness to submit one's actions to moral inquiry and be responsive to the assessment of others, which includes being accountable for meeting specific obligations, that is, liable to justify (or give reasonable excuses) the decisions, actions or means, and outcomes (sometimes unexpected), when required by the stakeholders or by law. The conflict between causal influence by the employer and moral responsibility of the employee is quite common in professions. In the engineering practice, the problems are:
     (a) The fragmentation of work in a project certainly makes the final products lie away from the immediate workplace, and lessens the personal responsibility of the employee.
     (b) The responsibilities diffuse into various hierarchies and to various people. Nobody gets the real feel of personal responsibility.
     (c) Often projects are executed one after another. An employee is more interested in adherence to tight schedules rather than giving personal care for the current project.
     (d) More legal action is to be faced by the engineers (as in the case of medical practitioners). This makes them wary of showing moral concerns beyond what is prescribed by the institutions.

Despite all these shortcomings, engineers are expected to face the risk and show up personal responsibility as the profession demands.

4.6   Balanced outlook on law

The 'balanced outlook on law' in engineering practice stresses the necessity of laws and regulations and also their limitations in directing and controlling the engineering practice. Laws are necessary because, people are not fully responsible by themselves and because of the competitive nature of the free enterprise, which does not encourage moral initiatives. Laws are needed to provide a minimum level of compliance.

Code for builders by Hammurabi [1]

Hammurabi, the king of Babylon, in 1758 framed the following code for the builders:

'If a builder has built a house for a man and has not made his work sound and the house which he has built has fallen down and caused the death of the householder, that builder shall be put to death. If it causes the death of the householder's son, they shall put that builder's son to death. If it causes the death of the householder's slave, he shall give slave for slave to the householder. If it destroys property, he shall replace anything it has destroyed; and because he has not made the house sound which he has built and it has fallen down, he shall rebuild the house which has fallen down from his own property. If a builder has built a house for a man and does not make his work perfect and the wall bulges, that builder shall put that wall in sound condition at his own cost.'

This code was expected to put self-regulation seriously in those years.

Steamboat code in the USA [1]

In the early 19th century, a law was passed in the USA to provide for inspection of the safety of boilers and engines in ships. It was amended many times and now the standards formulated by the American Society of Mechanical Engineers are followed.

4.7   Significance of law

Law is required to tackle a crisis. Whenever there is a fire accident in a factory or fire cracker's storehouse or boat capsize we make this claim, and soon forget. Laws are meant to be interpreted for minimal compliance. On the other hand, laws when amended or updated continuously would be counterproductive. Laws will always lag behind technological development. The regulatory or inspection agencies such as the Environmental authority of any country can play a major role by framing rules and enforcing compliance. Good laws when imposed effectively produce benefits. They establish minimum standards of professional conduct and provide an impetus to people. Further, they serve as moral support and defence for the people who are willing to act ethically.

It can be inferred that:

1. The rules that administer engineering practice should be construed as responsible experimentation rather than rules of a game. This makes the engineer responsible for the safe conduct of the experiment.

2. Precise rules and sanctions are suitable in case of ethical misconduct that involves the violation of established engineering procedures, which are aimed at the safety and the welfare of the public.

3. In situations where the experimentation is large and time-consuming, the rules must not try to cover all possible outcomes, and they should not compel the engineers to follow rigid courses of action.

4. The regulation should be broad, but make engineers accountable for their decisions, and

5. Through their professional societies, the engineers can facilitate framing the rules, amend wherever necessary, and enforce them, but without givingin for conflicts of interest.

4.8   Cautious optimism

A feeling of general self-belief regarding a situation and/or its result tied with readiness for feasible difficulties or failure. It's best to work out cautious optimism when starting any new business.

4.9   Safety and risk

Safety is important in any engineering discipline. One of the main duties of an engineer is to ensure the safety of the people who will be affected by the design of the products. The code of ethics of professional engineering societies stressed the value of safety. The engineering codes of ethics show that engineers have a responsibility to society to produce safe products. Nothing can be 100% safe, but engineers are required to make products as safe as reasonably possible. Thus safety should be an integral part of any engineering design. What may be safe for one person may not be safe for another person. A Power Saw in the hands of a child is unsafe, but it is safe in the hand of an adult. A sick adult is more prone to ill effects from air pollution than a healthy adult. What is safe to entrepreneurs may not be so to engineers,for example, Pilots: 'Indian Airports are not safe; Low Vision in Fog'. What is safe to engineers may not be so for the public. Typically several groups of people are involved in safety matters but have their interests at stake. Each group may differ in what is safe and what is not. 'A ship in harbour is safe, but that is not what ships are built for'. 'A thing is safe if its risks are judged to be acceptable'. 'A thing is safe (to a certain degree) with respect to a given person or group at a given time if they were fully aware of its risks and expressing their most settled values, they would judge those risks to be acceptable (to that certain degree)' (3).

Safe operation of the system and the prevention of natural or human-caused disasters.

Ex 1: We judge fluoride in water can kill lots of people. Here, we are overestimating risk.

Ex 2: We hire a taxi, without thinking about its safety. In that case, we are not estimating risk. A thing is not safe if it exposes us to unacceptable danger or hazard.

Risk in technology could include dangers of bodily harm, economic loss or environmental degradation and a situation involving exposure to danger. Absolute safety is never been possible. Any improvement in making a product safe involves an increase in the cost of production. The manufacturer and the user must have some understanding to know about the risk connected with any product and know how much it will cost to reduce those risks. Risk is the potential that something unwanted and harmful may occur. We take a risk when we undertake something or use a product that is not safe. Types of risk include acceptable risk, voluntary risk and control and job-related risks.

Acceptable risk refers to the level of human and property injury or loss from an industrial process that is considered to be tolerable by an individual, household, group, organization, community, region, state or nation in view of the social, political and economic cost–benefit analysis. For instance, the risk of flooding can be accepted once every 500 years but it is not acceptable every ten years. It is management's responsibility to set their company's level of risk. As a security professional, it is his/her responsibility to work with management and help them understand what it means to define an acceptable level of risk. Each company has its tolerance in risk level, which is derived from its legal and regulatory compliance responsibilities.

A person is said to take 'voluntary risk' when he is subjected to risk by either his own actions or action taken by others and volunteers to take that risk without any apprehension. Voluntary risks have to do with choices of everyday life. They are the risks that people take knowing that they may have consequences. These risks include smoking tobacco, driving a car, skydiving and climbing a ladder. Involuntary risks are risks that people take either not knowing that they are at risk, or they are unable to control the fact that they are at risk. These risks often include environmental hazards such as lightning, tsunamis and tornadoes.

Job-related risk: Many workers are taking risks in their jobs in their stride like being exposed to asbestos. Exposure to risks on a job is in one sense of voluntary nature since one can always refuse to submit to the work or may have control over how the job is done. But generally, workers have no choice other than what they are told to do since they want to stick to the only job available to them. But they are not generally informed about the exposure to toxic substances and other dangers which are not readily seen, smelt, heard or otherwise sensed. Occupational health and safety regulations and unions can have a better say in correcting these situations but still, things are far below expected safety standards.

4.9.1   Assessment of safety and risk

Absolute safety is never possible to attain and safety can be improved in an engineering product only with an increase in cost. On the other hand, unsafe products increase secondary costs to the producer beyond the primary (production)costs, like warranty costs loss of goodwill, loss of customers, legal action costs, downtime costs in manufacturing, etc. It should now be clear that 'safety comes with a price' only.

4.9.2   Goal of risk assessment

The risk assessment process aims to remove a hazard or reduce the level of its risk by adding precautions or control measures, as necessary. By doing so, one can create a safer and healthier workplace.

4.9.3   Safe exit

In the study of safety, the 'safe exit' principles are recommended. The conditions referred to as 'safe exit' are (4):

1. The product, when it fails, should fail safely.
2. The product, when it fails, can be abandoned safely (it does not harm others by explosion or radiation).
3. The user can safely escape the product (e.g., ships need a sufficient number of lifeboats for all passengers and crew; multi-storied buildings need usable fire escapes).

4.10   The Challenger case study

The NASA space shuttle Challenger exploded on 28 January 1986, just 73 s after liftoff, bringing a devastating end to the spacecraft's 10th mission (Fig. 4.1). The disaster claimed the lives of all seven astronauts aboard. It was later determined that two rubber O-rings, which had been designed to separate the sections of the rocket booster, had failed due to cold temperatures on the morning of the launch. The tragedy and its aftermath received extensive media coverage and prompted NASA to temporarily suspend all shuttle missions [5].

Figure 4.1 The challenger disaster

The safety lessons one can learn in the Challenger case are as follows:
1. Negligence in design efforts. The booster rocket casing recovered from earlier flights indicated the failure of filed-joint seals. No design changes were incorporated. Instead of two O-rings, three rings should have been fixed. But there was no time for testing with three rings. At least three rings could have been tried while launching.
2. Tests on O-rings should have been conducted down to the expected ambient temperature, i.e., 20°F. No normalization of deviances should have been allowed.
3. NASA was not willing to wait for the weather to improve. The weather was not favourable on the day of the launch. A strong wind shear might have caused the rupture of the weakened O rings.

4. The final decision-making of launch or no-launch should have been with the engineers and not with the managers. Engineers insisted on 'safety' but the managers went ahead with the 'schedule'.

5. Informed consent: The mission was full of dangers. The astronauts should have been informed of the probable failure of the O-rings (field joints). No informed consent was obtained, when the engineers had expressed that the specific launch was unsafe.

6. Conflict of interest (Risk vs. Cost): There were 700 criticality-1 items, which included the field joints. A failure in any one of them would have caused the tragedy. No backup or standby had been provided for these criticality-1 components.

7. Escape mechanism or 'safe exit' should have been incorporated in the craft. McDonnell Douglas, the engineer, designed an abort module to allow the separation of the orbiter, which was triggered by a field-joint leak. Unfortunately, such a safe exit was rejected due to the increase in the cost, simultaneously with a reduction in payload.

8. Ethical engineers should have been given awards and encouraged to hold their discretion (moral autonomy) in risky situations and to report to the appropriate agency their views, in the interest of public safety.

4.11   Bhopal gas tragedy

The cumulative effects of the following factors caused the tragedy in Bhopal on 3 December 1984 (Fig. 4.2).

Figure 4.2 Bhopal gas tragedy

1. Maintenance was neglected and the trained maintenance personnel were reduced as an economic measure. The need for quick diagnosis aggravates the situation by causing considerable psychological stress on the plant personnel.

2. Training activities for the supervisory personnel were stopped. This led to inadequate training of the personnel to handle emergencies.

3. Periodical Safety Inspection teams from the U.S., which visited previously were also stopped. From the initial U.S. Standards, the safety procedures were reduced to low-level Indian standards. The procedures had been deteriorating at these sites for weeks or months, before the accident. There was a clear lack of management systems and procedures to ensure safety.

4. Vital spares for equipment and machineries were not available.

5. Absence of capital replacement led to the stagnant economy of the plant.

6. The high turnover of the experienced engineers and technicians, who were demoralized by the lack of development.

7. Lack of experienced personnel to operate and control the vital installations.

8. They have not conducted a thorough process hazards analysis that would have exposed the serious hazards which resulted in disaster later.

9. No emergency plan was put in practice, during the shutdown and maintenance.

10. Above all, the commitment of top-level management to safety was lacking. They have been paying only lip service to the safety of the people of the host country.

Technologically, the tragedy was caused by a series of events listed below:

1. The safety manual of Union Carbide prescribed that the MIC tanks were to be filled only up to 60% of the capacity. But the tanks were reported to have been filled up to 75%.

2. The safety policy prescribed that an empty tank should be available as a stand-by in case of emergency. But the emergency tank was also filled to its full capacity. These facts confirmed that the MNC had not followed and implemented appropriate safety standards of the home country in the host country.

3. The storage tanks should be refrigerated to make the chemical less reactive. But here the refrigeration system was shut down as an economic measure. This raised the temperature of the gas stored.

4. The plant was shut down for maintenance two months earlier. The worker who cleaned the pipes and filters connected to the tanks and closed the valves was not trained properly. He did not insert the safety disks to prevent any possible leakage of the gas. This led to the build-up of temperature and pressure in the storage tanks.

5. When the gas started leaking out, the operators tried to use the vent gas scrubber that was designed to reduce the exhausting gas. But that scrubber was also shut down.

6. There was a flare tower that was designed to burn off the gas escaping from the scrubber. That was not also in working condition.

7. The workers finally tried to spray water up to 100 feet to quench the gas (which is water-soluble). But the gas was escaping from the chimney of 120-feet high.

8. The workers were not trained on safety drills or emergency drills or any evacuation plans.

The gas escaped into the air and spread over 40 km². About 600 people died and left 7000 injured and the health of about 2 million people has affected adversely. Even after 22 years, the influence of the Central Government and the courts, the compensation had not reached all the affected people.

Bhopal tragedy has taught us that every decision has a consequence [6]. In the case of Bhopal, the decision to turn off the safety systems turned out to cause more destruction than it should have. If the safety systems were been in working order, the leak would not have been so damaging. The second lesson that can be learned is that the legal system does not always work to protect victims. Bhopal has shown that sometimes corporations do not have to pay for the crimes they commit. The polluter does not always have to pay. The main lesson to be learned from the Bhopal Gas Disaster is about the dangers of the chemicals used everyday and the effect of those chemicals on human health and environment.

4.12   The Three Mile Island (TMI)

Three Mile Island is the site of a nuclear power plant in south-central Pennsylvania. On 28 March 1979, a series of mechanical and human errors at the plant caused the worst commercial nuclear accident in U.S. history, resulting in a partial meltdown that released dangerous radioactive gasses into the atmosphere (Fig. 4.3). Three Mile Island stoked public fears about nuclear power– no new nuclear power plants have been built in the United States since the accident.

Figure 4.3 The three Mile Island (TMI) nuclear disaster

4.12.1   Lessons learned from the incident [3]

1. Following the TMI accident, there was a belated push to complete the loss of flow and loss of coolant testing program that the Atomic Energy Commission had initiated in the early 1960s. For a variety of political, financial and managerial reasons, that program had received low priority and was chronically underfunded and behind schedule.

2. Today's plant designs undergo far more rigorous testing programs and have better, more completely validated computer models.

3. Far more attention has been focused on the possible impact of events like 'small break' loss of cooling accidents.

4. All new operators at pressurized water reactors learn to understand the importance of the relationship between saturation pressure and saturation temperature.

5. After the accident, the industry invested a great deal of effort into a sustained program to share the operating experience. The plant

designers also did not do their operators any favours in the design and layout of the control room. Key indicators were haphazardly arranged, there were thousands of different parameters that could cause an alarm if out of their normal range, and there was no prioritization of alarming conditions.

6. After the accident, an extensive effort was made to improve the control rooms for existing plants and to devise regulations that increased the attention paid to human factors, man-machine interfaces, and other facets of control room design. All plants now have their simulators that are designed to mimic the particular plant and are provided with the same operating procedures used in the actual plant. Operators are on a shift routine that puts them in the simulator for a week at a time every four to six weeks.

7. The initiating failures that started the whole sequence took place in the steam plant, a portion of the power plant that was not subject to as much regulatory or design scrutiny as the portions that were more closely associated with the nuclear reactor and its direct cooling systems.

8. An increased level of attention is now paid to structures, systems and components that are not directly related to a reactor, but there is still a confusing, expensive and potentially vulnerable system that attempts to classify systems and give them an appropriate level of attention.

9. Although most decision-makers in the nuclear industry understand the importance of planned maintenance systems to keep their equipment in excellent condition and the importance of a systematic approach to training to keep their employees performing at the best, they have not yet implemented an effective, adequately resourced, planned communications program that helps to ensure that the public and the media understand the importance of a strong nuclear energy sector.

4.13   Chernobyl disaster

The Chernobyl disaster, also referred to as the Chernobyl accident, was a catastrophic nuclear accident (Fig. 4.4). It occurred on 25–26 April 1986 in the No. 4 light water graphite moderated reactor at the Chernobyl Nuclear Power Plant near the now-abandoned town of Pripyat, in northern Ukrainian Soviet Socialist Republic, Soviet Union, approximately 104 km (65 miles) north of Kyiv.

Figure 4.4 Chernobyl disaster

Here are ten lessons drawn from the Chernobyl disaster [7]

1. Nuclear power is a highly complex, expensive and dangerous way to boil water. Nuclear power does nothing except provide a high-tech and extremely dangerous way to boil water to create steam to turn turbines.

2. Accidents happen and the worst-case scenario often turns out to be worse than imagined or planned for. Although the nuclear industry continues to assure the public that nuclear power plants are safe, the plants continue to have accidents, some of which exceed worst-case projections.

3. The nuclear industry and its experts cannot plan for every contingency or prevent every disaster. Although it was known that Fukushima

is subject to earthquakes and tsunamis, the nuclear industry and its experts did not plan for the combination of a 9.0 earthquake and the larger-than-expected tsunami that followed.

4. Governments do not effectively regulate the nuclear industry to assure the safety of the public. Government regulators of a nuclear industry often come from the nuclear industry and tend to be too close to the industry to regulate it effectively.

5. Hubris, complacency and high-level radiation are a deadly mix. Hubris on the part of the nuclear industry and its government regulators, along with complacency on the part of the public, have led to the creation of vast amounts of high-level radiation that must be guarded against release to the environment for tens of thousands of years, far longer than civilization has existed.

6. Nuclear power plants can catastrophically fail, causing vast human and environmental damage. The corporations that run the power plants, however, are protected from catastrophic economic failure by government limits on liability, which shift the economic burden to the public. If the corporations that own nuclear power plants had to bear the burden of potential financial losses in the event of a catastrophic accident, they would not build the plants because they know the risks are unacceptable. It is government liability limits, such as the Price-Anderson Act in the U.S., that make nuclear power plants possible, leaving the taxpayers responsible for the overwhelming monetary costs of nuclear industry failures. No other private industry is given such liability protection.

7. Radiation releases from nuclear accidents cannot be contained in space and will not stop at national borders. The wind will carry long-lived radioactive materials around the world and affect the people and environment of many countries and regions. The radiation will also affect the oceans of the world, which are the common heritage of humankind.

8. Radiation releases from nuclear accidents cannot be contained in time and will adversely affect countless future generations. The radioactive materials from nuclear power plant accidents, as well as from radioactive wastes, are a legacy we are bequeathing to future generations of humans and other forms of life on the planet.

9. Nuclear energy, as well as nuclear weapons, and human beings cannot co-exist without the risk of future catastrophes. The survivors of the

atomic bombings of Hiroshima and Nagasaki have long known that nuclear weapons and human beings cannot co-exist. The Fukushima accident, like that at Chernobyl before it, makes clear that human beings and nuclear power plants also cannot co-exist without courting future disasters.

10. The accident at Chernobyl is a wake-up call to phase out nuclear energy and replace it with energy conservation and more human- and environmentally-friendly forms of renewable energy. For decades it has been clear that various forms of renewable energy are needed to replace both nuclear and fossil fuel energy sources. Now it is clearer than ever. The choice is not between nuclear and fossil fuels. The solution is to disavow both of these forms of energy and to move as rapidly as possible to a global energy plan based upon various forms of renewable energy: solar cells, wind, geothermal, ocean thermal, currents, tides, etc.

References

1. Mike W. Martin and Schinzinger Roland, *Ethics in Engineering – Third Edition* (New Delhi: Tata McGraw-Hill Publishing Company Limited, 2003), 81, 114–115.

2. Rod A, 'What did we learn from TMI? March 25, 2014', Available from: http://ansnuclearcafe.org/2014/03/25/what-did-we-learn-from-three-mile-island/#sthash.3drM69b8.dpbs[Accessed July 30, 2018].

3. Mike W. Martin and Schinzinger Roland, *Ethics in Engineering – Second Edition* (New Delhi: Tata McGraw-Hill Publishing Company Limited, 2000), 132–133.

4. Naagarazan RS, *Professional ethics and human values* (New Delhi: Newage International Publishers, 2006), 48–49, 66.

5. Anon, 'Challenger Explosion', Available from:https://www.history.com/topics/challenger-disaster[Accessed July 29, 2018].

6. Anon, 'Bhopal Gas Disaster: 25 years of agony', Available from: https://bhopalgasdisaster.weebly.com/what-can-be-learned.html[Accessed July 29, 2018].

7. David K, 'Ten lessons from Chernobyl and Fukushima, July 15, 2016', Available from: https://www.wagingpeace.org/ten-lessons-chernobyl-fukushima [Accessed July 30, 2018].

5

Workplace rights and responsibilities

Abstract: Human rights are defined as moral entitlements that place obligations on other people to treat one with dignity and respect. Organizations and engineers are to be familiar with the minimum provisions under human rights so that the engineers and organizations constitute a firm base for understanding their role while achieving the productivity target. The chapter first discusses employee rights, that is, the moral and legal rights obtained with employee status, professional rights, basic human rights, institutional rights and non-contractual employee rights. The chapter then discusses teamwork, collegiality and loyalty, managing conflict, respect for authority, collective bargaining, confidentiality, conflicts of interest and occupational crime.

Keywords: professional rights; collegiality; loyalty; conflict of interest; collective bargaining; occupational crime

5.1   Introduction

The kind of commitments shown by the engineers understandably ranks high on the list of expectations that employers have from the engineers they employ or engage as consultants. Engineers in turn should attain top performance at a professional level as their main responsibility, accompanied by others such as maintaining the confidentiality and avoiding conflicts of interest. Engineers also need the opportunity to perform responsibly, and this means that their professional and employee rights must be respected.

Human rights are defined as moral entitlements that place obligations on other people to treat one with dignity and respect. Organizations and engineers are to be familiar with the minimum provisions under human rights so that the engineers and organizations for a firm base for understanding and productivity. Thomas Donaldson, Professor in Ethics and Law at the Wharton School at the University of Pennsylvania, formulates a list of 'international rights', human rights that are implied by, but more specific than, the most abstract human rights to liberty and fairness. Donaldson suggested 10 such international rights [1], which have been explained in Chap. 6.

5.2   Professional rights

Under professional rights, the following provisions are protected [2]:

1. Right to form and express professional judgment: It is also called the right of professional conscience. Pursuing professional responsibilities empowers one to form and exercise professional judgment. Both technical and moral judgments are included. This right is bound by the responsibilities of employers and colleagues.

2. Right to refuse to participate in unethical activities: It is also called the right of conscientious refusal. It is the right to refuse to engage in unethical actions and to refuse to do so solely because one view that as unethical. The employer cannot force or threaten the employee to do something that is considered by that employee as unethical or unacceptable. For example, unethical and illegal activities that can be refused are: falsifying data, forging documents, altering test results, lying, giving or taking bribes, etc. There may be situations when there is a disagreement or no shared agreement among reasonable people over whether an act is unethical. Medical practitioners have a right not to participate in abortions. Similarly, the engineers must have a right to refuse assignments that violate their consciences, such as when there exists a threat to human life or moral disagreement among reasonable people.

3. Right to fair recognition and to receive remuneration for professional services: Engineers have a right to professional recognition for their work and achievements. This includes fair monetary and non-monetary forms of recognition. It is related to morality as well as self-interest. They motivate them to concentrate their energy on jobs and to update their knowledge and skills through continuing education. This will prevent the engineers from diversion such as moonlighting or bother on money matters. Many times, the engineers who have contributed to getting patents on the organizations are not adequately remunerated. Based on the resources of the organization and the bargaining power of the engineers, the reasonable salary or remuneration for patent discovery can be worked out.

5.3   Employee rights

*   They are any rights, moral or legal, that involve the status of being an employee.

- They overlap with some professional rights, of the sort just discussed, and they also include institutional rights created by organizational policies or employment agreements, such as the right to be paid the salary specified in one's contract [3].
- They include some professional rights that apply to the employer–employee relationship.
- Employee rights include fundamental human rights relevant to the employment situation, for example, the right not to be discriminated against one's race, sex, age or national origin.

5.4   Right to privacy [2]

- It is the right to control the access to and use of information about oneself.
- It is limited in certain situations by employers' rights.
- Only duly authorized persons can get the personal information.
- A supervisor might suspect a worker and search his cupboard when the worker is absent. But the supervisor is to have another officer as a witness in such cases.

5.5   Right to choose outside activities [2]

- It means the right to have a private life outside the job.
- There are some situations when this right can be curbed, such as:
- When those activities lead to a violation.
- When moonlighting.
- When the interest of the employer is getting damaged.

5.6   Right to due process from the employer [2]

- It is the right to fair process or procedures in firing, demotion and in taking any disciplinary actions against employees.
- Written explanation should be initially obtained from the charged employee and the orders are given in writing with clearly stated reasons.
- Fairness here is specified in terms of the process rather than the outcomes.

5.7   Teamwork

Teamwork virtues are those that are especially important in enabling professionals to work successfully with other people. They include collegiality, cooperativeness, loyalty, and respect for legitimate authority [3]. Teamwork in the workplace and teamwork slogans are important in building morale and increasing productivity and loyalty. Employees in almost every workplace talk about 'their team', 'building the best team' and 'working as an efficient team', but very few understand what creating effective teamwork in the workplace entails. When one belongs to a team, one feels a part of belonging to something bigger than oneself. This includes understanding the aims and objectives of the company. A good and efficient team contributes to the success of the organization. When you work in an environment that is 'team-oriented' you automatically produce better results. Team building in a workplace needs competence. Good teamwork emerges when the team feels that all its members have the skill, knowledge and capability to handle issues or have the necessary access to all the help needed to accomplish the mission the team was created for. To have successful teamwork in the workplace the team has to have empowerment and freedom to feel accountable and responsible towards its vision and mission to accomplish it. Limitations and boundaries have to be set and understood by all team members in regards to how far they may go in their pursuit of answers/solutions. Therefore, time resources and monetary limitations should be defined in advance. Teamwork is the creation of a working culture that places 'collaboration' in high esteem. People in this type of environment understand and fully believe that planning, thinking, actions and decisions are far better when done in cooperation with one another. Values of effective teamwork in the workplace should be shared and identified with the employees. Rewards like bonuses, compensation and others should be given depending on joint teamwork as well as individual achievement and contribution.

5.8   Ethical corporate climate

An ethical climate is a working environment that is conducive to morally responsible conduct. Within corporations, it is produced by a combination of formal organization and policies, informal traditions and practices, and personal attitudes and commitments [3]. Engineers can make a vital contribution to such a climate, especially as they move into technical management and then more general management positions. Professionalism in engineering would be threatened at every turn in a corporation devoted primarily to powerful egos. Sociologist Jackall [4] describes several such corporations asorganizations that reduce (and distort) corporate values,that is, right in the corporation is what the guy above wants. That's what morality is in the corporation. Jackall describes a world in which professional standards are disregarded by top-level managers preoccupied with maintaining self-promoting images and forming powerful alliances with other managers. Hard work, commitment to worthwhile and safe products, and even profit-making take a back seat to personal survival in the tumultuous world of corporate takeovers and layoffs. Jackall's study was based on chemical and textile companies during the 1980s, Companies notorious for indifference to worker safety (including cotton-dust poisoning) and environmental degradation (especially chemical pollution).

What are the defining features of an ethical corporate climate?

There are at least four [3].

First, ethical values in their full complexity are widely acknowledged and appreciated by managers and employees alike. Responsibilities to all constituencies of the corporation are affirmed – not only to stockholders but also to customers, employees and all other stakeholders in the corporation.

Second, the use of ethical language is honestly applied and recognized as a legitimate part of a corporate dialogue. One way to emphasize this legitimacy is to make prominent a corporate code of ethics. Another way is to explicitly include a statement of ethical responsibilities in the job descriptions of all layers of management.

Third, top management sets a moral tone in words, in policies, and by personal example. Official pronouncements asserting the importance of professional conduct in all areas of the corporation must be backed by support for professionals who work according to the guidelines outlined in professional codes of ethics. Whether or not there are periodic workshops on ethics or formal brochures on social responsibility distributed to all employees, what is most important is fostering confidence that management is serious about ethics.

Fourth, there are procedures for conflict resolution. One avenue is to create ombudsersons or designated executives with whom employees can have confidential discussions about moral concerns. Equally important is educating managers about conflict resolution. There are also ties of loyalty and collegiality that help minimize conflicts in the first place.

The ethical culture in an organization can be thought of as a slice of the overall organizational culture. So, if the organizational culture represents 'how we do things around here,'the ethical culture represents 'how we do things around here in relation to ethics and ethical behaviour in the organization.'The ethical culture represents the organization's 'ethics personality.'

5.9   Collegiality

Collegiality is the cooperative relationship of colleagues. It is the tendency to support and cooperate with colleagues. It is a virtue essential for teamwork to be effective. According to NSPE, collegiality should include the following characteristics [5]:

- Engineers should not attempt to injure, unkindly or falsely directly or indirectly, the professional reputation, prospects, practice or employment of other engineers.
- Engineers should not untruthfully criticize other engineer's work.
- Engineers should bring the unethical or illegal practice of other engineers to the proper authority for action

According to Craig Ihara [5] – 'A kind of connectedness grounded in respect for professional expertise and commitment to the goals and values of the profession'.

Elements of collegiality are respect, commitment and connectedness.

1. Respect: In general, means valuing one's colleague for their professional skill and their devotion to the social goods promoted by the profession. For engineering, it means affirming the worth of other engineers engaged in producing socially useful and safe products.
2. Commitment: It is sharing a devotion to the moral ideals essential in the practice of engineering. Even when there is cut-throat competition between engineers, there should be a feeling that all engineers share a concern for the overall good to society.
3. Connectedness: It is the awareness of being part of a cooperative undertaking created by sharing commitments and skills. It means the sense of utility among engineers that includes cooperation and mutual support.

5.9.1   Why collegiality is a virtue?

Collegiality should be encouraged among engineers and other professionals because from the point of view of society, collegiality is the influential value to promote the aims of professions. From the point of professionals, collegiality is more valuable as many individuals jointly working for the goodness of the public and society.

5.9.2   Negative aspects of collegiality

Collegiality may be misused and distorted. For instance, colleagues appeal to be silent about corporate corruption. It may degenerate more groups of self-interest, rather than shared devotion to the public. Because of heavy competition among engineers, collegiality may focus on the corporate goal of maximizing profit at the expense of the public good.

5.10   Loyalty

Loyalty is the quality of being true and faithful in one's support. It is more a function of attitudes, emotions and a sense of identity. Senses of loyalty can be classified into agency loyalty and identification loyalty or attitude loyalty [2].

Agency loyalty is fulfilling one's prescribed duties to an employer. The contractual duties may include a particular task for which one is paid, general activities of cooperating with colleagues and following lawful authority with the organization. It concerns the matter of actions, whatever its motives. It is motivated by identification with the group to which one is loyal. Example: People may not like the job they do hate their employer, but still they would perform their duty as long as they are employees. This sense of loyalty is agency loyalty.

Identification loyalty or attitude loyalty is much concerned with attitudes, emotions, and a sense of personal identity as it does with action. Employees should meet their moral duties to the organization willingly with personal attachment and affirmation. The attitude of loyalty is more a virtue than an obligation. This type of loyalty is all right when the organizations work for the productivity or development of the community. Working together in the falsification of records or serious harm to the public does not merit loyalty. Further, with frequent takeovers or mergers resulting in a large-scale layoff, employees find it difficult to maintain attitude loyalty. The attitude loyalty is more a virtue than an obligation. This type of loyalty is all right when the organizations work for the productivity or development of the community. Working together in the falsification of records or serious harm to the public does not merit loyalty. Further, with frequent takeovers or mergers resulting in a large-scale layoff, employees find it difficult to maintain attitude loyalty.

5.10.1   Is loyalty obligatory (responsibility)?

Agency loyalty to employers is an obligation within proper limits. According to John H. Fielder, identification loyalty is obligatory, only when the two conditions are met. Employees must be treated fairly; they should be given their share of benefits and burdens. Employees must see that their goals are achieved by and through a group in which they participate. Identification loyalty is reciprocal. That is employees can be expected to be loyal to employers only when employers show strong commitments to them.

5.10.2   Professionalism and loyalty

Acting on professional commitments to the public is more effective to serve a company than just following company orders. Loyalty to employers may not mean obeying one's, immediate supervisor. Professional obligations to both an employer and to the public might strengthen rather than contradict each other.

5.11   Managing conflict/conflict of interest

In general conflicts of interest mean individuals as two or more desires that all interests cannot be satisfied given circumstance. Professional conflicts of interest are situations where professionals,who have an interest, if pursued, could keep from meeting one of their obligations to their employers,for example,employees working in a company serving as a consultant for a competitor's company.

Types of conflict of interest are of three types [2]. They are actual conflicts of interest, potential conflicts of interest and apparent conflicts of interest. Actual conflicts of interest refer to a situation where objectivity is lost in decision-making and the inability to discharge the duty to the employer. Apparent conflicts of interest lead to doubting the engineer's interest and the ability for professional judgment. Potential conflicts of interest arise in situations where the interest of an employee extends beyond the current employer and into the interest of one's spouse, relative or friend. The interest changes into intimacy and subsequent non-moral judgments against the interest of the employer and in favour of the outsider or even a potential competitor. Examples of potential conflicts of interest can be a favourable contract, bribe and gift, moonlighting and insider information.

Favourable contract is a situation when an engineer's spouse is working for a contractor or a vendor, a conflict does not arise. But if the engineer is to give a subcontract to the contractor or purchase order to the supplier, the conflict arises.

In the case of bribes and gifts, the conflict arises when accepting large gifts from the suppliers. A bribe is different from a gift. Codes of ethics do not encourage even gifts, but employees have set forth flexible policies. Government and company policies generally ban gifts of more than a nominal value. An additional thumb rule is that the acceptance of a gift should not influence one's judgment on merit.

Moonlighting [2] is a situation when a person is working as an employee for two different companies in the spare time. This is against the right to pursue one's legitimate self interest It will lead to a conflict of interests if the person works for competitors, suppliers or customers while working under an employer. Another effect of moonlighting is that it leaves the person exhausted and harms the job performance in both places.

Insider Information is another potential conflict of interest is when using 'inside' information to establish a business venture or get an advantage for oneself or one's family or friends. The information may be either of the parent company or its clients or its business partners, for example, engineers might inform the decision on the company's merger with another company or acquisition or an innovative strategy adopted. In such cases, their friends get information on stock holding and decide on trading their stocks to sell or buy quickly, so that gain more or prevent a loss.

5.11.1   How to avoid conflicts of interest?

One can take guidance from company policy. In the absence of such a policy, taking a second opinion from a co-worker or manager is an option. This gives an impression that there no intention on the part of the engineer to hide anything.

5.12   Respect for authority

Respect for authority is important in meeting organizational goals. Decisions must be made in situations where allowing everyone to exercise unrestrained individual discretion would create chaos. Moreover, clear lines of authority provide a means for identifying areas of personal responsibility and accountability [3]. It is the right to make decisions, the right to direct the work, and the right to give orders. It is a crucial factor in an organization since engineers and employees must be authorized to carry out the jobs assigned to them. Authority can be defined as the legal right to command action by others to enforce compliance. Clear lines of authority identify areas of personal responsibility and accountability. Authority derives from several sources. They are the person's position or rank, and personal attitudes, such as charisma, knowledge and expertise. Institutional authority can be defined as the institutional right given to a person to exercise power based on the resource of the institution. It is an authority given by the institution to the qualified individuals to meet their industries objectives. This authority is exercised by making policy, allocating resources, issuing orders, carrying out actions, giving recommendations, etc. Limitations include that it is given by owners. In practice sometimes, it is given to ineffective persons. They are unable to exercise their authority effectively to meet the company's objectives.

Experts' authority [2]: It is the possession of special knowledge skills, competencies to perform some task or to give sound advice. It proved that leaders with expertise can effectively guide and motivate others than conventional leaders. This concept is referred to as 'authority of leadership'. In today's organization, the staff engineers, advisors, and consultants are given expert authority, while the institutional authority is assigned to the line managers.

There is a difference between authority and power. Authority is the legal right to a superior, which compels his subordinates to perform certain acts. Power is the ability of the person to influence others to perform an act. Authority is delegated to an individual by his supervisor. Power is earned by an individual by his effort. Authority is mostly well-defined and finite. Power is undefined and infinite. Authority lies in the position held and the authority change in position. Power resets in the individual. Even when their position has changed, his power remains with him.

Morally justified authority: The institutional authority assigned to the employee may ensure the achievement of the institutional objectives. But those institutional rights should necessarily be morally justified institutional rights and duties. The institutional authority is said to be morally justified when the goals of the institution are morally permissible or morally desirable and the way of implementation should not violate basic moral duties.

Accepting authority [2]: Employees accept their employers' authority by accepting the guidance and obeying the directives issued by the employer. According to Herbert Simon, 'a subordinate is said to accept authority whenever he permits his behaviour to be guided by the decision of superior, without independently examining the merits of that decision'. All the employers have the limits on 'zone of acceptance' in which they are willing to accept the authority. Generally, employees are not interested to make an issue of every incident of questionable morality, because of fear of losing their job. Therefore the 'zone of acceptance' can be used as a measure of the lack of individual moral integrity.

5.13   Collective bargaining [2]

International Labour Organization (ILO) defines it as negotiation about working conditions and terms of employment between employer and one or more representative employees with a view to reaching the agreement. The term bargaining refers to an evolving agreement using methods like negotiation, discussion, exchange of facts and ideas rather than confrontation. The process of collective bargaining includes presenting the character of demands by the union on behalf of constituent elements, compromise at the bargaining table and reaching the agreement

5.14   Unionism and professionalism

Legally, any organization employing more than 20 employees could have a union. In an organization, more than one union is permitted. The employees form unions to safeguard their interests and to prevent exploitation by employers. According to John Kemper, Dean, College of Engineering of the University of California,unionism and professionalism are conflicting with each other [5].

Professionalism offers paramount importance to the interest of society and their employers. But unions are collective bargaining agents. Many professional societies indirectly instruct the engineers should not become members of the unions. Collective bargaining is ethical or unethical only based on the given situation:

1. Arguments in favour of unions
It plays a vital role in achieving high salaries and an improved standard of living. Employees get a greater sense of participation in organization decisions and job security. Unions maintain stability by providing an effective grievance procedure for employee complaints. Unions can act as a counter force to any political movement that exploits the employees.

2. Arguments against unions
Unions destroy the economy of a country. Unions remove person-to-person negotiation between employers and employees. Unions encourage conflict

and stressed relations between employees and employers. Unions prevent employers from rewarding individuals for their achievements.

5.15   Confidentiality

Information considered desirable to be kept secret is called confidential [3]. Any information that the employer or client would like to have kept secret to compete effectively against business rivals is called confidentiality. This information includes how business is run, its products, and suppliers, which directly affects the ability of the company to compete in the marketplace.

Privileged information [2] is the information available only based on special privileges such as granted to an employee working on a special assignment.

Proprietary information [2] is that a company owns or is the proprietor of. This is primarily used in a legal sense and also called a trade secret. A trade secret can be virtually any type of information that has not become public and which an employer has taken steps to keep secret.

Patents differ from trade secrets in that they legally protected specific products from being manufactured and sold by competitors without the express permission of the patent holder. They have the drawback of being public and competitors may easily work around them by creating alternate designs.

5.15.1   Justification of confidentiality

Confidentiality can be justified by various ethical theories [2]. According to rights-based theory, the rights of the stakeholders, right to the intellectual property of the company are protected by this practice. Based on duty theory, employees and employers have to keep up mutual trust. The utilitarian theory holds good, only when confidentiality produces the most good to most people. Act utilitarian theory focuses on each situation when the employer decides on some matters as confidential.

Moral principles also justify the concept of confidentiality in the following context [2]:

I. Respect for autonomy:

It means respecting the freedom and self-determination of individuals and organizations to identify their legitimate control over the personal information of themselves. In the absence of this, they cannot keep their privacy and protect their self-interest.

II. Regard for public well-being:

Only when there is confidence that the physician will not reveal information, the patient will have the trust to confide in him. Similarly only when companies maintain some degree of confidentiality concerning their products, the benefits of competitiveness within a free market are promoted.

III. Respect for promise

This means giving respect for the promises made between employers and employees. Employees should not disclose the promises given to the employers. This information may be considered sensitive by the employer.

IV. Trustworthiness

Employees are obliged to protect confidential information regarding former employment, after a change of job. The confidentiality trust between employer and employee continues beyond the period of employment.

5.16   Occupational crime

Occupational crimes are illegal acts made possible through one's lawful employment [2]. It is a secretive violation of laws regulating work activities. When committed by office workers or professionals, occupational crime is called 'white-collar crime'. Most occupational crimes are special instances of conflicts of interest. These crimes are motivated by personal greed, corporate ambition and misguided company loyalty. Some of the examples are as follows [2]:

I. Price fixing

While fixing the price for any product or service sometimes all competitors come together and jointly set up the price to be charged. These are called pricing cartels. This is an unfair and unethical practice.

II. Endangering lives

Some companies employ workers without disclosing harmful health effects and safety hazards about the working environment. This is a kind of occupational crime.

III. Industrial espionage

Espionage refers secret gathering of information to influence relationships between two entities. The vital information is secretly gathered through espionage agents for economic gains.

IV. Bootlegging

Manufacturing, selling and transporting products (liquor and narcotics) that are prohibited by law, is called bootlegging. In the engineering context, it refers to working on projects which are prohibited or not properly authorised.

People who are committing occupational crime usually have a high standard of education and from the non-criminal family background. Middle-class males around 27 years of age (70% of the time) with no previous history, no involvement in drug or alcohol abuse and those who had troublesome life experiences in childhood are usually involved in this crime.

References

1. Donaldson Thomas, *The Ethics of International Business* (New York: Oxford University Press, 1989), 81.

2. Naagarazan R.S., *Professional ethics and human values* (New Delhi: Newage International Publishers, 2006), 71–72, 74–79, 80–81, 149–150.

3. Mike W. Martin and Schinzinger Roland, *Ethics in Engineering – Second Edition* (New Delhi: Tata McGraw-Hill Publishing Company Limited, 2000), 61, 132–133, 143–144, 146–147, 150.

4. Jackall Robert, 'Moral Mazes: The World of Corporate Managers', *International Journal of Politics, Culture, and Society* 1, no. 4, (Summer, 1988): 598–614.

5. Anon, 'Responsibilities and rights', Available from: https://srikarthiks.files.wordpress.com/2016/02/unit-4.ppt[Accessed August 17, 2018].

6

Global issues

Abstract: Globalization means amalgamation of countries through commerce, transfer of technology and exchange of information and culture. The increasing international pour of capital, technology, trade and people had persuaded to change the nature of local organizations, governments and people which resulted in social changes and development. The chapter discusses issues such as multinational organizations, computers, internet functions, military development and environmental ethics, which have embedded greater importance for their very sustenance and advancement for the engineers. While dealing with varieties of responsibilities under different situations, the chapter then discusses the role of engineers transforms into a manager, consultant, expert witness and advisors. Moral leadership which presents the engineers with many challenges to their moral principles is essential to encourage different working groups to achieve morally enviable goals. The reason why such moral leadership is required for engineers has also been explained in this chapter.

Keywords: international rights; environmental ethics; sentient-centred ethics; bio-centric ethics; hacking; moral leadership

6.1   Multinational corporations

Organizations that have reputable businesses in more than one country are called multinational corporations. The headquarters are in the home country and the business is extended in many host countries. The Western organizations doing business in the less-economically developed (developing and overpopulated) countries gain the advantage of inexpensive labour, availability of natural resources, conducive-tax atmosphere and virgin market for the products. At the same time, the developing countries are also benefited by fresh job opportunities, jobs with higher remuneration and challenges, transfer of technology and several social benefits by the wealth developed. But this happens invariably with some social and cultural disturbance. Loss of jobs for the home country, and loss or exploitation of natural resources, political instability for the host countries are some of the threats of globalization [1].

6.2   Technology transfer

It is a process of moving technology to a new setting and implementing it there. Technology includes hardware (machines and installations) and techniques (technical, organizational and managerial skills and procedures). It may mean moving the technology applications from the laboratory to the field/factory or from one country to another. This transfer is affected by governments, organizations, universities, and MNCs [1].

6.3   Appropriate technology

Identification, transfer and implementation of the most suitable technology for a set of new situations, is called appropriate technology. Technology includes both hardware (machines and installations) and software (technical, organizational and managerial skills and procedures). Factors, such as economic, social and engineering constraints are the causes for the modification of technology. Depending on the availability of resources, physical conditions (such as temperature, humidity, salinity, geographical location, isolated land area and availability of water), capital opportunity costs and the human value system (social acceptability), which includes their traditions, beliefs and religion, the appropriateness is to be determined. For example, small farmers in our country like to own and use the power tillers, rather than the high-powered tractors or sophisticated harvesting machines. On the other hand, the latest technological devices, cell phones and wireless local loop phones have found their way into remote villages and hamlets than the landline telephone connections. Large aqua-culture farms should not make existing fishermen jobless in their village. The term appropriate is value-based and it should ensure fulfilment of the human needs and protection of the environment [1].

6.4   International human rights

The moral responsibilities and obligations of the multinational corporations operating in the host countries are dependent on the framework of rights ethics. Common minimal rights are to be followed to rationalise the transactions when the engineers and employers of MNCs have to interact at official, social, economic and sometimes political levels. At the international level, the organizations are expected to adopt the minimum levels of (a) values, such as mutual support, loyalty and reciprocity, (b) the negative duty of refraining from harmful actions such as violence and fraud and (c) basic fairness and practical justice in case of conflicts.

Thomas Donaldson, Professor in Ethics and Law at the Wharton School at the University of Pennsylvania, formulates a list of international rights. These international rights have great importance and are often put at risk. Their exact requirements must be understood contextually, depending on the traditions and economic resources available in particular societies. International rights proposed by Donaldson are 10 in numbers [2].

The ten international rights to be taken care of, in this context are:

1. Right of freedom of physical movement of people.
2. Right of ownership of properties.
3. Freedom from torture.
4. Right to a fair trial on the products.
5. Freedom from discrimination based on race or sex. If such discrimination against women or minorities is prevalent in the host country, the MNC will be compelled to accept. MNCs may opt to quit that country if the human rights violations are severe.
6. Physical security. Uses of safety gadgets have to be supplied to the workers even if the laws of the host country do not suggest such measures.
7. Freedom of speech and forming an association.
8. Right to have minimum education.
9. Right to political participation.
10. Right to live and exist (i.e., co-existence). The individual liberty and sanctity of human life are to be respected by all societies.

These are human rights; as such they place restrictions on how multinational corporations can act in other societies, even when those societies do not recognize the rights in their laws and customs. For example, the right to non-discriminatory treatment would make it wrong for corporations to participate in discrimination against women and racial minorities even though this may be a dominant custom in the host country. Again, the right to physical security requires supplying protective goggles to workers running metal lathes, even when this is not required by the laws of the host country.

6.5   Promoting morally just measures

Richard T. De George, University distinguished Professor of philosophy, of Russian and East European studies, and business administration, and co-director of the International Centre for Ethics in Business at the University of Kansas, agrees that multinational corporations should respect the basic rights of people in the countries where they do business, but he requires more, especially when wealthy countries do business in less economically developed countries. In the spirit of utilitarianism, which calls for promoting the most good for the most people, he also requires that the activities of multinational corporations benefit the host countries in which they do business [3]. The business activities of multinational corporations must do more overall good than bad, which means helping the country's overall economy and its workers, rather than benefiting a few corrupt leaders in oppressive regimes. Not only must they pay their fair share of taxes, but they must make sure the products they manufacture or distribute are not causing easily preventable harms. Also, the overall impact of the business dealings must tend to promote morally just institutions in the society, not increase unjust institutions. At the same time, corporations should respect the laws and culture of the host country, providing they do not violate basic moral rights. Of course, there is a tension between promoting just institutions and respecting local customs. For example, as U.S. business attempts to encourage a move toward treating women equally in the workplace, they may undermine local customs, often supported by religion, about the appropriate roles for women. Good judgment exercised in good faith, rather than abstract principles, is often the only way to address such practical dilemmas.

De George calls for a contextual, case-by-case approach in applying principles of human rights and promoting the good of the host country. If multinational corporations pay exactly the pay rate of the host country, they will be accused of exploiting workers, especially when that rate is below a living wage sufficient for the person to live with dignity as a human being. If they pay well beyond that rate they will be accused of unfairly stealing the most skilled workers in the society, drawing them away from other companies important to the local economy. De George's guideline is to pay a living wage, even when local companies fail to pay such a wage, but otherwise, pay only enough to attract competent workers. As another example, consider the issue of worker safety in companies that manufacture hazardous chemicals. When is it permissible for the U.S. to transfer dangerous technology such as asbestos production to another country and thensimply adopt that country's safety laws? Workers have the right to informed consent. Even if the host country does not recognize that right, corporations are required to inform workers, in a language they can understand, of the dangers. Workers may be so desperate for income to feed their families that they will work under almost any conditions. Corporations must eliminate great risks when they can while still making a reasonable profit. They must also pay workers for the extra risks they undertake as indicative of morally good judgment and negotiation.

6.6   Environmental ethics

Environmental ethics is the study of (a) moral issues concerning the environment, and (b) moral perspectives, beliefs or attitudes concerning those issues [1]. Engineers in the past are known for their negligence of the environment, in their activities. It has become important now that engineers design eco-friendly tools, machines, sustainable products,processes, and projects. These are essential now to (a) ensure protection (safety) of the environment, (b) prevent the degradation of the environment and (c) slow down the exploitation of the natural resources, so that the future generation can survive. The American Society of Civil Engineers (ASCE) code of ethics, has specifically required that 'engineers shall hold paramount the safety, health, and welfare of the public and shall strive to comply with the principles of sustainable development in the performance of professional duties'

The term sustainable development emphasizes the investment, orientation of technology, development and functioning of organizations to meet the present needs of people and at the same time ensuring the future generations to meet their needs.

Engineers as experimenters have certain duties towards environmental ethics [1], namely:

1. Environmental impact assessment: One major but sure and unintended effect of technology is wastage and the resulting pollution of land, water, air and even space. Study how the industry and technology affect the environment.

2. Establish standards: Study and fix the tolerable and actual pollution levels.

3. Counter measures: Study what protective or eliminating measures are available for immediate implementation.

4. Environmental awareness: Study how to educate the people on environmental practices, issues and possible remedies.

6.7   Role of engineers in sustainability

The group of people that maintains, enhances or improves its environmental, social, cultural and economic resources in such a way that support current and future community members in pursuing healthy, productive and happy lives can very well be termed as Sustainable Community. Professional engineers play an important and significant role to meet sustainability [4]. They work to improve the welfare, health and safety, with the minimal use of natural resources and paying attention with regard to the environment and the sustainability of resources. Sustainability is influenced by challenges and opportunities. Options and solutions to maximize social value and minimize environmental impacts are to be provided by engineers. There are some grave challenges because of the undesirable effects of exhaustion of resources, rapid population growth and damage the ecosystems and environmental pollution. Only an environmentally sociable advancement is not sufficient and increasingly engineers are required to take a wider perspective including societal integrity, local, universal associations and poverty mitigation. Comprehensiveness brings crucial prospects for engineers to promote change through sharing experiences and through good quality practice. The authoritative responsibility and guidance of engineers in achieving sustainability should not be underestimated. Increasingly this will be part of multidisciplinary teams that include non-engineers and through work that crosses national boundaries.

The main goal of sustainable development is to enable people throughout the world to meet and satisfy their basic needs and enjoy an improved and better quality of life without compromising the quality of life for generations to come. Sustainable development has largely being categorized in two perceptions, requirements and boundaries imposed by the state of technology and the social organization on the environment's ability to meet present and future demands. Following principles have been established to achieve sustainable development.

- Living within the environmental goals.
- Ensuring a strong, healthy and justified society.
- Promotion of good governance.
- Achieving a sustainable and efficient economy with the conscientious use of science.
- Responsible to maximize the value of their activity to build a sustainable planet.
- Empathies the important potential role for engineering.
- Empathies about the environmental limits and finite resources.
- Reduce the demand for resources.

- Reduction of waste production by using effectively the resources that are used.
- Make use of systems and products which reduce embedded carbon, energy and water use, waste and pollution, etc.
- Adoption of full life-cycle assessment as normal practice including the supply chain.
- Adopt strategies, such as salvaging, reprocessing, decommissioning and discarding of components and materials.
- During the design stage itself minimization of any adverse impacts on sustainability.
- Carrying out a comprehensive risk assessment prior to starting of the project.
- Risk assessment should ensure and includes the potential environmental, economic and societal impacts.
- Monitoring systems to measure any environmental, social and economic impacts of engineering projects so they can be identified at an early stage.

6.8   Engineers responsibilities towards environment

Scientific research continues to provide information about the links between human health and environmental quality. Essential components of life are air, water and food, which provide potential pathways for contaminants to affect our health. Air, water and soil pollutions exposure has been linked to various diseases/disorders like cancer, lupus, immune diseases, allergies and asthma, problems in reproduction and birth defects, allergic reactions, nervous system disorders, hypersensitivity and decreased resistance to diseases [4].

(a) Air pollution

Air pollution is a great threat to our sustainable environment. Engineers in every country of the world should try to

- cut down the release of sulphur dioxide, nitrogen oxide, carbon dioxide and mercury through regulatory programs according to established targets and time frames.
- involve themselves in national and international initiatives to address transboundary air issues.

- work to meet standards for two primary components of smog(formed mainly above urban centres, is composed mainly of troposphere ozone (O_3) ground-level ozone and particulate matter.
- build up air-shed management plans and team up with large industrial facilities to monitor transport and deposition from major sources.

(b) Water pollution

Safe drinking water is another challenge for many developing countries where engineers in the world can contribute a good deal to this issue. So, engineers should try to

- resolve quality and quantity issues of water for agriculture and fisheries sectors and develop a scaffold for safeguard water resources and aquatic habitat that builds on the drinking water strategy.
- consult with the Municipal/Public Works Association of the relevant region while developing guidelines, standards and regulations for issues related to municipal water and wastewater.
- employ a government-wide approach to water problems through the Interdepartmental Drinking Water Management Committee.
- develop, modify and upgrade the ambient water monitoring system with proper maintenance.
- team-up with the Department of Health to tackle issues related to contaminants in drinking water.
- address wastewater issues by working with municipal and domestic partners.
- be a support system to municipalities for their water and wastewater infrastructure programs and land use planning in water supply areas.

(c) Land pollution

Harmful substances in water, air and soil cause noteworthy health perils. The concerned government is devoted to minimise the environmental impacts of such materials and protecting the country's health. In this regard, engineers will

- efficiently promote pollution prevention.
- validate risk-based administrative approaches to spotlight efforts where they are most needed.
- bring up-to-date existing directives controlling hazardous substances and eliminating regulatory duplication.

- promote effective utilisation, storage, handling and discarding of harmful substances.
- apply the 'polluter pays' principle to users of hazardous substances.
- encourage stewardship by manufacturers to promote proper life cycle management of hazardous substances.
- make joint efforts with other authorities to perk up treatment of contaminated sites and promote sustainable redevelopment.
- promote early detection and response to land quality issues through legislated requirements for mandatory reporting of site contamination.

6.9   Engineering, ecology and economics

Like the word ethics, the expression of environmental ethics can have several meanings. We use the expression to refer tothe study of moral issues concerning the environment and moral perspectives, beliefs and attitudes concerning those issues.

The invisible hand and the commons.

Two metaphors have dominated thinking about the environment: the invisible hand and the tragedy of the commons.

1st metaphor

This addresses the positive side in the economic and environmental aspects. The invisible hand serves as the motivational drive for a person to get their interest.

2nd metaphor

This takes account of the damage to the environment. People tend to be thoughtless about things they do not own individually and which seem to be an unlimited supply.

6.9.1   Engineers: sustainable development

- There is a 'standard engineering worldwide that of mechanical universe' which is at odds with mainstream 'organic' environmental thought. Central of the engineering view is 'techno think' which implicitly assumes that things can be understood by analysing them and, if something goes wrong, can be fixed.
- Green philosophy demands humility, respect and sensitivity towards the natural world.

- The engineering code of ethics explicitly refers to environmental responsibilities under the heading of sustainable development. Engineers should be committed to improving the environment to change the quality of life.

6.9.1.1   Sustainable development

The term is used in a negative way to underscore how current patterns of economic activity and growth cannot be sustained as the population grows, technologies are extended to developing countries, and the environment is increasingly harmed. On a positive front, the term implies the crucial need for new economic patterns and products that are sustainable, that are compatible with both ongoing technological development and protection of the environment.

6.9.2   Anthropocentric environmental ethics

Anthropocentric environmental ethics focuses exclusively on the benefits of the natural environment to humans and the threats to human beings presented by the destruction of nature. This approach assumes that only human beings have inherent moral worth duly to be taken care of. Other living beings and ecosystems are only instrumental. Utilitarianism aims to maximize good consequences for human beings. Most of the goods are engineered products made out of natural resources. Human beings have also (a) recreational interests (enjoy leisure through mountaineering, sports and pastimes), (b) aesthetic interests (enjoy nature as from seeing waterfalls and snow-clad mountains), (c) scientific interests to explore nature or processes and (d) a basic interest to survive, by preservation as well as conservation of nature and natural resources. Rights ethicists favour the basic rights to live and the right to liberty, to realise the right to live in a supportive environment. Further, virtue ethics stresses the importance of prudence, humility, appreciation of natural beauty and gratitude to nature.

However, nature-centred ethics, which ensures the worth of all living beings and organisms, seems to be more appropriate in the current context. Many Asian religions stress the unity with nature, rather than domination and exploitation. Zen Buddhism calls for a simple life with compassion towards humans and other animals. Hinduism enshrines the ideal of oneness (Advaita) and the principle of ahimsa to all living beings. It identifies all human beings, animals and plants as divine. The eco-balance is the need of the hour and the engineers are the right experimenters to achieve this.

6.9.3   Sentientcentred ethics

One version of nature-centred ethics recognises all sentient animals as having inherent worth. Sentient animals are those that feel pain and pleasure and have desires. Thus, some utilitarians extend their theory (that right action maximises goodness for all affected) to sentient animals as well as humans. Peter Singer, an Australian moral philosopher, Professor of Bioethics at Princeton University and a Laureate Professor at the Centre for Applied Philosophy and Public Ethics at the University of Melbourne, developed a revised act-utilitarian perspective in his influential book, 'Animal Liberation'. Singer insists that moral judgments must take into account the effects of our actions on sentient animals. Failure to do so is a form of discrimination akin to racism and sexism. He labels it as

'speciesism: a prejudice or attitude of bias toward the interests of members of one's species and against those of members of other species'[5].

In Singer's view, animals deserve equal consideration, in that their interests should be weighed fairly, but that does not mean equal treatment with humans (as their interests are different from human interests). Thus, in building a dam that will cause flooding to grasslands, engineers should take into account the impact on animals living there. Singer allows that sometimes animals' interests have to give way to human interests, but their interests should always be considered and weighed.

The singer does not ascribe rights to animals, and hence it is somewhat ironic that Animal Liberation has been called the bible of the animal rights movement. Other philosophers, however, do ascribe rights to animals. Tom Regan, an American philosopher who specialized in animal rights theory and professor emeritus of philosophy at North Carolina State University,contends that conscious creatures have inherent worth not only because they can feel pleasure and pain, but because more generally they are subjects of experiences who form beliefs, memories, intentions, and preferences and they can act purposefully[6]. In his view, their status as subjects of experiments makes them sufficiently like humans to give them rights. Singer and Regan tend to think of inherent worth as all or nothing. Hence they think of conscious animals as deserving equal consideration. That does not mean they must be treated in the identical way we treat humans, but only that their interests should be weighed equally with human interests in making decisions. Other sentient ethicists disagree. They regard conscious animals as having inherent worth, although not equal to that of humans [7].

6.9.4   Bio-centric ethics

Life-centred ethics regards all living organisms as having inherent worth. Albert Schweitzer (1875–1965), an Alsatian theologian, organist, writer, humanitarian, philosopher, and physician, set forth a pioneering version of this perspective under the name of 'reverence for life'. He argued that our most fundamental feature is not our intellect but instead our will to live, by which he meant both a will to survive and a will to develop according to our innate tendencies. All organisms share these instinctive tendencies to survive and develop, and hence consistency requires that we affirm the inherent worth of all life. More than an appeal to logical consistency, however, Schweitzer appealed to what has been called 'bio-empathy',that is,our capacity to experience a kinship with other life, to experience other life in its struggle to survive and grow. Empathy, if we allow it to emerge, grows into sympathy and compassion, gradually leading us to accept

'as good preserving life, promoting life, developing all life that is capable of development to its highest possible value' [8].

Schweitzer often spoke of reverence for life as the fundamental excellence of character, and hence his view is a version of nature-centred virtue ethics. He refused to rank forms of life according to degrees of inherent worth, but he believed that a sincere effort to live by the ideal and virtue of reverence for life would enable us to make inevitable decisions about when life must be maintained or has to be sacrificed. More recent defenders of bio-centric ethics, however, have developed complex sets of rules for guiding decisions.

Paul Taylor, an American philosopher, for example, provides extensive discussion of four duties: (1) nonmaleficence, which is the duty not to kill other living things; (2) noninterference, which is the duty not to interfere with the freedom of living organisms; (3) fidelity, which is the duty not to violate the trust of wild animals (as in trapping) and (4) restitution, which is the duty to make amends for violating the previous three duties [9]. These are prima facie duties, which have exceptions when they conflict with overriding moral duties and rights, such as self-defence.

6.9.5   Eco-centric ethics

A frequent criticism of sentient-centred and bio-centred ethics is that they are too individualistic, in that they locate inherent worth in individual organisms. Eco-centred ethics locate inherent value in ecological systems. This more holistic approach was voiced by the naturalist Aldo Leopold (1887–1948), who urged that we must promote the health of ecosystems. In one of the most famous statements in the environmental literature, he wrote:

'A thing is right when it tends to preserve the integrity, stability, and beauty of the biotic community. It is wrong when it tends otherwise' [10].

This 'land ethic,'as he called it, implied a direct moral imperative to preserve (leave unchanged), not just conserve (use prudently), the environment, and to live with a sense that we are part of nature, rather than that nature is a mere resource for satisfying our desires. More recent defenders of eco-centric ethics have included within this holistic perspective an appreciation of human relationships. Thus, J. Baird Callicott, an American philosopher, writes that

'an eco-centric ethic does not replace or cancel previous socially generated human-oriented duties to family and family members, to neighbours and neighbourhood, to all human beings and humanity'[11].

That is, locating inherent worth in wider ecological systems does not cancel out or make less important what we owe to human beings.

6.10   Computer ethics and internet

Computers have become the technological backbone of society. Their degree of complexity, range of applications, and sheer numbers continue to increase. Computers raise a host of difficult moral issues, many of them connected with basic moral concerns such as free speech, privacy, respect for property, informed consent and harm [12]. To evaluate and deal with these issues, a new area of applied ethics called computer ethics has sprung up. Computer ethics has special importance for the new groups of professionals emerging with computer technology, for example, designers of computers, programmers, systems analysts, and operators. To the extent that engineers design, manufacture, and apply computers, computer ethics is a branch of engineering ethics. But many professionals who use and control computers share the responsibility for their applications.

6.10.1   Types of issues

Different types of problems are found in computer ethics.

1. Computer as the instrument of unethical acts
(a) The usage of computers replaces the job positions: this has been overcome to a large extent by readjusting work assignments and training everyone on computer applications, such as word processing, editing and graphics. (b)

Breaking privacy: information or data of the individuals accessed or erased or the ownership changed. (c) Defraud a bank or a client: by accessing and withdrawing money from another's bank account.

2. Computer as the object of unethical act

The data are accessed and deleted or changed.

(a) Hacking: the software is stolen or information is accessed from other computers. This may cause financial loss to the business or violation of the privacy rights of the individuals or business. In case of defence information being hacked, this may endanger the security of the nation. (b) Spreading virus: through mail or otherwise, other computers are accessed and the files are erased or contents changed altogether. 'Trojan horses' are implanted to distort the messages and files beyond recovery. This again causes financial loss or mental torture to the individuals. Some hackers feel that they have justified their right to free information or they do it for fun. However, these acts are certainly unethical. (c) Health hazard: the computers pose threat during their use as well as during disposal.

3. Problems related to the autonomous nature of computer

(a) Security risk: once the Tokyo Stock Exchange faced a major embarrassment. A seemingly casual mistake by a junior trader of a large security house led to huge losses including that of reputation. The order through the exchange's trading system was to sell one share for 600,000 Yen. Instead, the trader keyed in a sale order for 600,000 shares at the rate of one Yen each. Naturally, the shares on offer at the ridiculously low price were lapped up. And only a few buyers agreed to reverse the deal. The loss to the securities firm was said to be huge, running into several hundred thousand. More important to note, such an obvious mistake could not be corrected by some of the advanced technology available. For advanced countries like Japan who have imbibed the latest technology, this would be a new kind of learning experience [13].

(b) Loss of human lives: risk and loss of human lives lost by computer, in the operational control of military weapons. There is a dangerous instability in the automated defence system. An unexpected error in the software or hardware or a conflict during interfacing between the two may trigger a serious attack and cause irreparable human loss before the error is traced. The Chinese embassy was bombed by the U.S. military in Iraq a few years back, but enquiries revealed that the building was shown in a previous map as the building where insurgents stayed. (c) For flexible manufacturing systems, the autonomous computer is beneficial in obtaining continuous monitoring and automatic control.

Various issues related to computer ethics are discussed as follows:

The ethical problems initiated by computers in the workplace are:

1. Elimination of routine and manual jobs:this leads to unemployment, but the creation of skilled and IT-enabled service jobs are more advantageous for the people. Initially, this may require some up-gradation of their skills and knowledge, but formal training will set this problem right. For example, in place of a typist, we have a programmer or an accountant.

2. Health and safety: the ill-effects due to electromagnetic radiation, especially on women and pregnant employees, mental stress, wrist problem known as Carpel Tunnel Syndrome, and back pain due to poor ergonomic seating designs, and eye strain due to poor lighting and flickers in the display and long exposure, have been reported worldwide. Throughout long exposure, these are expected to affect the health and safety of the people. The computer designers should take care of these aspects and management should monitor the health and safety of the computer personnel.

3. Computer failure: failure in computers may be due to errors in the hardware or software. Hardware errors are rare and they can be solved easily and quickly. But software errors are very serious as they can stop the entire network. Testing and quality systems for software have gained relevance and importance in the recent past, to avoid or minimize these errors.

6.10.2   Property issues

The property issues concerned with the computers are [1]:

1. Computers have been used to extort money through anonymous telephone calls.

2. Computers are used to cheat and steal by current as well as previous employees.

3. Cheating and stealing from the customers and clients.

4. Violation of contracts on computer sales and services.

5. Conspiracy as a group, especially with the internet, defrauds the gullible, stealing the identity, and forge documents.

6. Violation of property rights: is the software a property? The software could be either a program (an algorithm, indicating the steps in solving

a problem) or a source code (the algorithm in a general computer language such as FORTRAN, C and COBOL or an object code (to translate the source code into the machine language). How do we apply the concept of property here? This demands a framework for ethical judgments.

Property is what the laws permit and defines as can be owned, exchanged and used. The computer hardware (product) is protected by patents. The software (idea, expression) is protected by copyrights and trade secrets. But algorithms cannot be copyrighted, because the mathematical formulas can be discovered but not owned. The object codes which are not intelligible to human beings cannot be copyrighted. Thus, we see that reproducing multiple copies from one copy of (licensed) software and distribution or sales are crimes. The open-source concepts have, to a great extent, liberalized and promoted the use of computer programs for the betterment of society.

6.10.3   Privacy and anonymity

The data transmission and accessibility have improved tremendously by using the computers, but the right to privacy has been threatened to a great extent. Some issues [1] concerned with privacy are listed here:

1. Records of evidence

Service records or criminal records and the details of people can be stored and accessed to prove innocence or guilt. Records on psychiatric treatment by medical practitioners or hospitals, or records of membership of organizations may sometimes embarrass the persons in later years.

2. Hacking

There are computer enthusiasts who wilfully or for fun, plant viruses or 'Trojan horses' that may fill the disc space, falsify information, erase files and even harm the hardware. They breakdown the functioning of computers and can be treated as a violation of property rights. Some hackers opine that the information should be freely available for everybody. It is prudent that the right to individual privacy in limiting the access to information on oneself should not be violated. Further any unauthorized use of personal information (which is property), is to be considered as theft. Besides individual privacy, national security and freedom within the economy are to be respected. The proprietary information and data of the organizations are to be protected so that they can pursue the goals without hindrance.

3. Legal response

In the Indian scene, the Right to Information Act 2005 [14] provides the right to the citizens to secure access to information under the control of public authorities, including the departments of the central government, state governments, government bodies, public sector companies and public sector banks, to promote transparency and accountability of public authorities.

Right to information: Under the Act, section 2 (j), the right to information includes the right to (1) Inspect works, documents, records, (2) take notes, extracts or certified copies of documents or records, (3) take certified samples of material and (4) obtain information in the form of printouts, diskettes, floppies, tapes, video cassettes or in any other electronic mode.

4. Anonymity

Anonymity in computer communication has some merits as well as demerits. While seeking medical or psychological counselling or discussion (chat) on topics, such as AIDS, abortion, gay rights, anonymity offers protection (against revealing their identity). But frequently, anonymity is misused by some people for money laundering, drug trafficking and preying upon the vulnerable.

6.11   Weapons development

Military activities including the world wars have stimulated the growth of technology. The growth of the Internet amply illustrates this fact. The development of warfare and the involvement of engineers bring out many ethical issues concerned with engineers, such as the issue of integrity in experiments as well as expenditure in defence research and development, issue of personal commitment and conscience, and the issues of social justice and social health. Engineers involved in weapons development because of the following reasons [1]:

1. It gives one job with a high salary.
2. One takes pride and honour in participating in the activities towards the defence of the nation.
3. One believes they fight a war on terrorism and thereby contribute to the peace and stability of the country. The wars have never won peace, only peace can win the peace.
4. By research and development, the engineer is reducing or eliminating the risk from enemy weapons, and saving one's country from disaster.

5. By building up arsenals and show of force, a country can force the rogue country, towards regulation. Engineers can participate effectively in arms control negotiations for surrender or peace.

6.12   Engineers as managers

Characteristics

The characteristics [1] of engineers as managers are:
1. Promote an ethical climate, through framing organization policies, responsibilities and personal attitudes and obligations.
2. Resolving conflicts, by evolving priority, developing mutual understanding, generating various alternative solutions to problems.
3. Social responsibility to stakeholders, customers and employers. They act to develop wealth as well as the welfare of society. Ethicists project the view that the manager's responsibility is only to increase the profit of the organization, and only the engineers have the responsibility to protect the safety, health and welfare of the public. But managers have the ethical responsibility to produce safe and good products (or useful service) while showing respect for human beings who include the employees, customers and the public. Hence, the objective for the managers and engineers is to produce valuable products that are also profitable.

6.13   Solving conflicts

In solving conflicts, force should not be resorted. The conflict situations should be tolerated, understood and resolved by participation by all the concerned. The conflicts in the case of project managers arise in the following manners [1]:
(a) Conflicts based on schedules: This happens because of various levels of execution, priority and limitations of each level.
(b) Conflicts arising out of fixing the priority to different projects or departments. This is to be arrived at from the end requirements and it may change from time to time.
(c) Conflict based on the availability of personnel.
(d) Conflict over technical, economic and time factors such as cost, time, and performance level.
(e) Conflict arising in administration such as authority, responsibility, accountability, and logistics required.

(f)   Conflicts of personality, human psychology and ego problems.

(g)   Conflict over expenditure and its deviations.

Most of the conflicts can be resolved by following the principles listed [1] here:

1. People

Detachment of people from the problem implies that the views of all concerned should be obtained. The questions such as what, why, and when the error was committed are more important than knowing who committed it. This impersonal approach will lead to not only early solutions but also others will be prevented from committing errors.

2. Interests

The focus must be only on interest, that is, the ethical attitudes or motives and not on the positions (i.e., stated views). A supplier may require a commission larger than the usual prevailing rate for an agricultural product. But the past analysis may tell us that the material is not cultivated regularly and the monsoon poses some additional risk to the supply. Mutual interests must be respected to a maximum level.

3. Options

Generate various options as solutions to the problem. This helps a manager to try the next best solution should the first one fails. Decisions on alternate solutions can be taken more easily and without loss of time.

4. Evaluation

The evaluation of the results should be based on some specified objectives, such as efficiency, quality and customer satisfaction. Thus, more important is that the means, not only the goals, should be ethical.

6.14   Consulting engineers

The consulting engineers work in private. There is no salary from the employers. But they charge fees from the sponsor and they have more freedom to decide on their projects. Still, they have no absolute freedom, because they need to earn their living. The consulting engineers have ethical responsibilities different from the salaried engineers, as follows [1]:

1. Advertising

The consulting engineers are directly responsible for advertising their services, even if they employ other consultants to assist them. But in many organizations, this responsibility is with the advertising executives and the personnel department. They are allowed to advertise but to avoid deceptive ones. Deceptive advertising such as the following are prohibited:

(a) By white lies, i.e.,a lie that is told to be polite or to stop someone from being upset by the truth.

(b) By half-truth, e.g., a product has been tested as a prototype, but it was claimed to have been already introduced in the market. An architect shows the photograph of the completed building with flowering trees around but actually, the foundation of the building has been completed and there is no real garden.

(c) Exaggerated claims, e.g.,the consultant might have played a small role in a well-known project. But they could claim to have played a major role.

(d) Making false suggestions. For instance, the cost reduction might have been achieved along with the reduction in strength, but the strength details are hidden.

(e) Through vague wording or slogans.

2. Competitive bidding

It means offering a price and gets something in return for the service offered. The organizations have a pool of engineers. The expertise can be shared and the bidding is made more realistic. But the individual consultants have to develop creative designs and build their reputation steadily and carefully, over a while. The clients will have to choose between the reputed organizations and proven qualifications of the company and the expertise of the consultants. Although competent, the younger consultants are thus slightly at a disadvantage.

3. Contingency fee

This is the fee or commission paid to the consultant when one is successful in saving the expenses for the client. A sense of honesty and fairness is required in fixing this fee. The NSPE Code III 6 (a) says that the engineers shall not propose or accept a commission on a contingent basis where their judgment may be compromised. The fee may be either as an agreed amount or a fixed percentage of the savings realized. But in the contingency fee agreements, the judgment of the consultant may be biased. The consultant may be tempted to specify inferior materials or design methods to cut the construction cost. This fee may motivate the consultants to effect saving in the costs to the clients, through reasonably moral and technological means.

4. Safety and client's needs

The greater freedom for the consulting engineers in decision making on safety aspects, and difficulties concerning truthfulness are the matters to be given attention to. For example, in design-only projects, the consulting engineers may design something and have no role in the construction. Sometimes, difficulties may crop up during construction due to the non-availability of suitable materials, some shortcuts in construction, and lack of necessary and adequate supervision and inspection. Properly trained supervision is needed, but may not happen unless it is provided. Further, the contractor may not understand and/or be willing to modify the original design to serve the clients best. A few on-site inspections by the consulting engineers will expose the deficiency in execution and save the workers, the public, and the environment that may be exposed to risk upon completion of the project.

The NSPE codes on the advertisement by consultants provide some specific regulations. The following are the activities prohibited in an advertisement by a consultant:

1. The use of a statement containing misrepresentation or omission of a necessary fact.
2. Statement intended or likely to create an unjustified expectation.
3. Statement containing prediction of future (probable) success.
4. Statement intended or likely to attract clients, by the use of slogans or sensational language format.

6.15   Engineers as expert witness

Frequently engineers are required to act as consultants and provide expert opinions and views in many legal cases of past events. They are required to explain the causes of accidents, malfunctions and other technological behaviour of structures, machines and instruments, e.g., personal injury while using an instrument, defective product, traffic accident, structure or building collapse, and damage to the property is some of the cases where testimonies are needed. The focus is on the past. The engineers, who act as expert-witnesses, are likely to abuse their positions in the following manners [1]:

1. Hired guns

Most lawyers hire engineers to serve the interest of their clients. Lawyers are permitted and required to project the case in a way favourable to their clients. But the engineers have obligations to thoroughly examine the events and demonstrate their professional integrity to testify only the truth in the court. They do not serve the clients of the lawyers directly. The hired guns forward white lies and distortions, as demanded by the lawyers. They even withhold the information or shade the fact, to favour their clients.

2. Money bias

Consultants may be influenced or prejudiced for monitory considerations, gain reputation and make a fortune.

3. Ego bias

The assumption that the own side is innocent and the other side is guilty, is responsible for this behaviour. An inordinate desire to serve one's client and get name and fame is another reason for this bias.

4. Sympathy bias

Sympathy for the victim on the opposite side may upset the testimony. The integrity of the consultants will keep these biases away from the justice. The court also must obtain a balanced view of both sides, by examining the expert witnesses of lawyers on both sides, to remove a probable bias.

Duties:

1. The expert witness is required to exhibit the responsibility of confidentiality just as they do in the consulting roles. They cannot divulge the findings of the investigation to the opposite side unless it is required by the court of law.

2. More important is that as a witness they are not required to volunteer evidence favourable to the opponent. They must answer questions truthfully, need not elaborate, and remain neutral until the details are asked for further.

3. They should be objective to discover the truth and communicate them honestly.

4. The stand of the experts depends on the shared understanding created within the society. The legal system should be respected and at the same time, they should act in conformance with the professional standards as obtained from the code of ethics.

5. The experts should earnestly be impartial in identifying and interpreting the observed data, recorded data, and the industrial standards. They should not distort the truth, even under pressure. Although they are hired by lawyers, they do not serve the lawyers or their clients. They

serve justice. Many a time, their objective judgments will help the lawyer to put up the best defence for their clients [1].

6.16   Engineers as advisors

6.16.1   Advisors

The engineers are required to give their view on the future such as in planning, policy-making, which involves technology. Various issues and requirements for engineers who act as advisors [1]are:

1. Objectivity

The engineers should study the cost and benefits of all possible alternative means objectively, within the specified conditions and assumptions.

2. Study all aspects

They have to study the economic viability (effectiveness), technical feasibility (efficiency), operational feasibility (skills) and social acceptability, which include environmental and ethical aspects, before formulating the policy.

3. Values

Engineers have to possess the qualities, such as (a) honesty, (b) competence (skills and expertise), (c) diligence (careful and alert), (d) loyalty in serving the interests of the clients and maintaining confidentiality and (e) public trust, and respect for the common good, rather than serving only the interests of the clients or the political interests.

4. Technical complexity

The arbitrary, unrealistic, and controversial assumptions made during future planning that are overlooked or not verified, will lead to moral complexity. The study on the future is full of uncertainties than the investigations on the past events. On the study of energy options, for example, assumptions on population increase, lifestyle, urbanization, availability of local fossil resources, projected costs of generating alternative forms of energy, world political scenario, world military tensions and pressures from world organizations such as World Trade Organization (W.T.O.) and European Union (EU) may increase the complexity in judgment on future.

5. National security

The proposed options should be aimed to strengthen the economy and security of the nation, besides safeguarding the natural resources and the environment from exploitation and degradation.

For the advisors on policy making or planning, a shared understanding on balancing the conflicting responsibilities, both to the clients and to the public, can be achieved by the following roles or models [1]:

1. Hired gun

The prime obligation is shown to the clients. The data and facts favourable to the clients are highlighted, and unfavourable aspects are hidden or treated as insignificant. The minimal level of interest is shown for public welfare.

2. Value-neutral analysts

This assumes an impartial engineer. They exhibit conscientious decisions, impartiality, i.e., without bias, fear or favour and absence of advocacy.

3. Value-guided advocates

The consulting engineers remain honest (frank in stating all the relevant facts and truthful in the interpretation of the facts) and autonomous (independent) in judgment and show paramount importance to the public (as different from the hired guns).

6.17   Moral leadership

Engineers provide many types of leadership in the development and implementation of technology, as managers, entrepreneurs, consultants, academics and officials of the government. Moral leadership is not merely the dominance of a group. It means adopting reasonable means to motivate the groups to achieve morally desirable goals. This leadership presents the engineers with many challenges to their moral principles.

Moral leadership is essentially required for the engineers, for the reasons listed [1] as follows:

1. It is leading a group of people towards the achievement of global criteria in leadership and related objectives. The goals, as well as the means, are to be moral. For example, Hitler and Stalin were leaders, but only in an instrumental sense and certainly not in a moral sense.

2. The leadership shall direct and motivate the group to move through morally desirable ways.

3. They lead by thinking ahead in time and morally creative towards new applications, extension and putting values into practice. 'Morally creative' means the identification of the most important values as applicable to the situation, bringing clarity within the groups through proper communication, and putting those values into practice.

4. They sustain professional interest, among social diversity and cross-disciplinary complexity. They contribute to professional societies, their professions, and their communities. Moral leadership in engineering is manifested in leadership within professional societies. The professional societies provide a forum for communication and canvassing for change within and by groups.

5. Voluntarism: This is another important avenue for providing moral leadership within communities, by the engineers is to promote services without fee or at reduced fees (pro bono) to the needy groups. The professional societies can also promote such activities among the engineers. This type of voluntarism (or philanthropy) has been in practice in the fields of medicine, law and education. But many of the engineers are not self-employed as in the case of physicians and lawyers. The business institutions are encouraged to contribute a percentage of their services as free or at concessional rates for charitable purposes.

6. Community service: This is another platform for engineers to exhibit their moral leadership. The engineers can help in guiding, organising, and stimulating the community towards morally- and environmentally desirable goals. The corporate organizations have come forward to adopt villages and execute many social welfare schemes, towards this objective. The Codes of Ethics promote and sustain the ethical environment and assist in achieving the ethical goals in the following manner:

   (a) It creates an environment in a profession, where ethical behaviour is the basic criterion.

   (b) It guides and reminds the person as to how to act, in any given situation.

   (c) It provides support to the individual, who is being pressurized or tortured by a superior or employer, to behave unethically.

   (c) Apart from professional societies, companies and universities have framed their own codes of ethics, based on the individual circumstances and specific mission of the organizations. These codes of conduct help in employees' awareness of ethical issues, establish and nurture a strong corporate ethical culture.

References

1. Naagarazan R.S, *Professional ethics and human values* (New Delhi: Newage International Publishers, 2006), 89, 91, 94, 95, 99, 100, 102–109.

2. Donaldson Thomas, *The Ethics of International Business*(New York: Oxford University Press, 1989), 81.

3. De George Richard T., *Ethical Dilemmas for Multinational Enterprise: A Philosophical Overview,in Ethics and the Multinational Enterprise*, ed. W. Michael Hoffman, Ann E. Lange, and David A. Fedo (New York: University Press of America, 1986), 4.

4. Dutta Amit Bijon, Sengupta Ipshita (2014), 'Engineering and sustainable environment', *International Journal of Engineering Research and General Science*, 2, 6, 124–130.

5. Singer Peter, *Animal Liberation*, rev. ed. (New York: Avon Books, 1990), 6.

6. Regan Tom, *The Case for Animal Rights*(Berkeley, CA: University of California Press, 1983).

7. Midgley Mary, *Animals and Why They Matter*(Athens, GA: University of Georgia Press, 1984).

8. Schweitzer Albert, *Out of My Life and Thought*, trans. A. B. Lemke (New York: Henry Holt and Company, 1990), 157. See also Mike W. Martin, Albert Schweitzer's *Reverence for Life* (Hampshire, UK: Ashgate Publishing, 2007).

9. Taylor Paul W., *Respect for Nature*, (Princeton, NJ: Princeton University Press, 1986).

10. Leopold Aldo, *A Sand County Almanac,*(New York: Ballantine, 1970), 262.

11. Callicott J. Baird, *Environmental Ethics, Encyclopedia of Ethics*, vol. 1, ed. L.C. Becker (New York: Garland, 1992), 313–314.

12. Bowie Norman E., *Ethical Issues in Information Technology*, The*Blackwell Guide to Business Ethics*, (Malden, MA: Blackwell, 2002), 267–288.

13. Narasimhan C.R.L., *Technology alone not to blame for risks*, (India: The Hindu, Feb. 20, 2006), Business review page.

14. Govt of India, 'Right to Information Act 2005', Available from: www.persmin.nic.in [Accessed August 8, 2018].

Index

A

Abuse 48
Accepting authority 82
Accountability 59
Accuracy 10
Agency loyalty 79, 80
Air pollution 93
Anonymity 103
Anthropocentric environmental
    ethics 96
Appropriate technology 88
Attitude 25
ASCE 91
ASME 37

B

Balanced outlook on law 60
Beliefs 4
Beneficence 44
Bhopal gas tragedy 65
Bio-centric ethics 98
Bootlegging 86
Breaking privacy 100
Broad outlook 58
Bureaucratic servant 40

C

Caring 2, 14
Cautious optimism 61

Challenger 64
Chernobyl disaster 66
Civic virtue 12
Code of ethics 46, 47
Collective bargaining 83
Collegiality 78, 79
Conceptual inquiry 27
Courage 2, 16
Cooperation 2, 17
Commitment 2, 17, 18, 78
Competitive bidding 106
Computer ethics and internet 99
Confidentiality 84
Conflicting reasons 29
Connectedness 78
Conflict of interest 64, 80, 81
Conventional 31, 32, 33
Conscientiousness 58
Consensus 34
Consulting engineers 105
Contingency fee 106
Controversy 34
Corporate social responsibility 46

D

Descriptive inquiry 28
Discernment 10
Duty based ethics 40, 43
Duty ethics 46
Discrimination 45

E

Eco-centric ethics 98, 99
Ecology 95
Economic rights 42
Egalitarian 34
Ego bias 108
Ethics 1,4, 5, 6, 7
Ethical corporate climate 76
Ethical dilemmas 29
Ethical theories 45
Empathy 2,19
Employee rights 85
Endangering lives 74,75
Engineering ethics 23
Engineering as experimentation 56
Engineers as advisors 108
Engineers as managers 104
Engineers as expert witness 107
Engineers as responsible
     experimenters 57
Environmental ethics 91
Experts' authority 82

F

Favoritism 45
Favourable contract 80
Factual inquiry 28
Fidelity 44, 98

G

Gilligan's theory 32, 33
Golden mean ethics 40

Gratitude 44
Green philosophy 95
Guardian 39

H

Hacking 100, 102
Human values 1, 10
Honesty 1, 7, 8
Hedonism 4
Hired guns 105, 110
Human rights 42, 73

I

IEEE 37
IEEE code of ethics 51
Industrial espionage 85
Inequality 48
Informed consent 64
Inquiry 27
Insider information 81
Institution of engineers' code of
     ethics 53
Integrity 1, 7, 8
International rights 89
International human rights 88
Intellectual courage 16
Involvement 17

J

Justice 44
Justice theory 45

K

Kohlberg's theory 30, 33

L

Land pollution 94
Legal response 103
Living peacefully 13
Loyalty 2, 79, 88
Love 9
Life centered ethics 98

M

Managing conflict 80
MIC 66
Military activities 103
Moonlighting 81
Money bias 108
Morals 1, 2, 4, 5, 6, 7
Moral values 1
Moral judgement 1
Morality 1
Moral awareness 1
Moral autonomy 30, 34
Moral reasoning 1
Moral imagination 1
Moral communication 1
Moral dilemmas 29
Moral issues 24
Moral leadership 110
Moral sovereignty 59
Morally justified authority 82
Multinational corporations 87

N

NASA 64
Non-interference 98

Non-violence 9
Normative inquiry 27
Non-maleficence 44, 98
NSPE 48, 106, 107
Nuclear power 68, 69

O

Occupational crime 85
Opportunity 24
O-rings 64

P

Peace 9
Perseverance 10, 17
Physical courage 16
Physical security 89
Post-conventional 31, 32, 33
Pre-conventional 31
Price fixing 85
Privileged information 84
Profession 36, 37
Professionalism 36, 37, 83
Professional obligations 50
Professional rights 74
Proprietary information 84
Privacy and anonymity 102

R

Respect 78
Respect for authority 81
Respect for autonomy 84
Respect for others 13
Respect for promise 85

Respect for public well-being 85

Resource crunch 24

Restitution 97

Reparation 44

Right based ethics 40, 41

Right to choose outside activities 75

Right to information 103

Right to privacy 75

Risk assessment 93

S

Safe exit 63, 65

Safety and risk 61, 63

Saviour 39

Security risk 100

Service learning 2, 20, 21

Self-confidence 2, 15

Self-improvement 44

Self-realization ethics 45

Sentient centered ethics 97

Sharing 2, 14

Social enabler 40

Social servant 40

Society 11

Social courage 16

Solving conflicts 104

Space shuttle 64

Speciesism 97

Spirituality 2, 19, 20

Stage theory 31

Sustainability 92

Sustainable community 92

Sustainable development 92, 95, 96

Sympathy bias 108

T

Team work 76

Technology transfer 88

Three-mile Island 67

Truth 9

Truthfulness 7

Trustworthiness 7, 85

U

Unenforceability 48

Unethical corporate behaviour 48

Union carbide 66

Unionism 83

Utilitarian ethics 40

Utilitarian theory 44, 45, 46

V

Values 2, 3, 4, 5, 8, 9, 10

Vagueness 29

Virtues 2

Virtue theory 41

W

Water pollution 94

Weapons development 103

White-collar crime 85

Work ethics 11

Rs 2195-co
$$\frac{7/12}{S\ \&\ T.}$$